Passive Solar: Subdivisions, Windows, Underground

Passive Solar:
Subdivisions, Windows, Underground

edited by
Herb Wade, Jeffrey Cook,
Ken Labs & Steve Selkowitz

American Solar Energy Society
Kansas City Chapter - Missouri Solar & Energy Association
1983

ACKNOWLEDGEMENTS

Herb Wade for major production work.

Brick Brewer of Brewer Energy Productions for carrying to completion.

K.C.-M.S.E.A. for sponsorship.

Billie Bledsoe for help on typesetting.

Ken Schmitz of Show-Me Printing for his help and the use of his shop.

Drew Henrickson for additional art.

NOTICE

Library of Congress Cataloging in Publication Data

Main entry under title:

Passive solar.

1. Solar Energy—Passive systems 2. Solar houses.
3. Earth sheltered houses. I. Wade, Herb.

TH7413.P372 1982 690'.869 82-16341

ISBN 0-89553-057-0

Table of Contents

SUBDIVISIONS (Edited by Herb Wade)

WINDOWS (Edited by Steve Selkowitz)

UNDERGROUND (Edited by Ken Labs)

THE COMMERCIAL PHOTO CO., INC. – BOB ASKREN

The first *Passive Solar Pioneers* award was presented to William Keck and George Fred Keck at the 4th National Passive Solar Conference in Kansas City, Missouri, by J. Douglas Balcomb (right).

Honoring George Frederick Keck and William Keck as Passive Solar Pioneers

Upon recommendation of the Northern Illinois Solar Energy Association, the Passive Division of the American Section of the International Solar Energy Society conveyed honors and recognition of George Frederick Keck and William Keck at Kansas City. These two brothers and their distinctive passive solar architectural practice were the feature at a special ceremony and presentation on 3 October. Recipients of numerous architectural awards over the years, they designed their first passive solar homes in the 1930's before most of the participants at passive solar conferences were born. This, however, is the first time they were honored for their solar leadership.

Keck and Keck had the first architectural office to professionally demonstrate the aesthetic potential of passive solar architecture. They were solar pioneers, not because they discovered some new or revolutionary thermal principle, but rather because they synthesized old ideas with new materials in an evolutionary way and created a substantial number of handsome and livable buildings. In a harsh climate, they designed in tune with nature. Now, over 40 years later, their early practice continues to inspire design quality for passive solar architecture.

In the December, 1979, issue of the American Institute of Architects *Journal* Allan Feeman wrote:

People have been "discovering" the Kecks for several years now. Their houses of the past 50 years are original, logical and ingenious. Many, not surprisingly, are beautiful. Although frequently published in the 1940s and '50s, their work drew less attention in the '60s and early '70s until the "Chicago Architects" exhibit of 1976. Now they are becoming deserved heroes of passive solar designers, many of whom are in their 20s and 30s and eager to trace their intellectual roots.

The evidence of their early passive concern is impressive. Solar orientation, earth berming, exterior Venetian blinds, roof ponds for cooling, heat sinks—all were used by the Kecks before World War II, although they are reluctant to take credit for ideas used by the ancient Chinese, Koreans and Romans. Ask them about their earliest influences; they will tell you about their father, a furniture retailer in Watertown, Wis., a man of German descent who made every dollar go as far as possible. Later, in the 1930s, "when we found we could save a little heat by organizing the windows on the south side, that sounded like a damn good idea," recalls William. During this period, Fred was principal designer and William was writing specifications. Not coincidentally, the modernist style in which the Kecks have always designed—Wright was a strong influence—was compatible with their "common sense" approach to all aspects of architecture, energy conservation included.

George Fred Keck first made a name for himself with two houses he designed and built for Chicago's "Century of Progress" World's Fair in 1933-34. (A 1920 architectural engineering graduate of the University of Illinois, Fred had opened his own practice in 1927 and was joined by William after the brother was graduated from U.I. in 1931. William became a full partner in 1946.) Neither the House of Tomorrow, a duodecagonal, center core structure with fixed plate glass facade, nor the Crystal House, also glass-sheathed but supported by exterior trusses running up the sides and over the roof, was solar designed. The point of these houses was prefabrication—quick construction, no excavation, low cost and maintenance—and the Kecks returned to prefabrication during World War II. But the winter heat gain through the glazing made an impression on the Kecks, and passive solar concepts showed up one by one in their houses after the fair. These ideas were reused "whenever possible, in luxury houses as well as the more modest ones," says William.

In 1935, they dug the north side of a house into a sloped river site at Watertown. A couple of years later, they put exterior operable Venetian blinds on the Fricker house in Whitewater, Wis., and on an apartment building in Chicago's Hyde Park, the building in which they live today. (The high sulfurous acid content in Chicago's air so corroded the stainless steel clips and chains that they were replaced by fixed, all-aluminum blinds in the 1940s. "Live and learn," says William.)

They put a cooling pond on the roof of a 1938 house, installing a spigot for the owner to operate when the water is depleted. And at about this time they began calculating solar angles, widening southern overhangs and putting black slate floors in sun rooms. "We didn't have the term 'heat sink'; we just said it was a nice place to keep your feet warm," recalls William, who also has a strong affinity for radiant flue tiles as flooring. The Kecks used these in pre-World War II years with both hot air and water. Radiant tiles provide even, draftless heat—"magnificent; like sitting in the sun, but without the glare," he says, lamenting that they were killed off by high labor installation costs and by the postwar demand for airconditioning, whose ducts make radiant systems redundant.

Subsequent work has built upon and refined their prewar passive solar efforts, and the firm is still designing houses around Chicago. The Kecks' ingenuity with the sun is typical of their pragmatic and humane approach to design.

The Development of Passive Solar Applications in the Midwest

George Fred Keck, William Keck, A.I.A.
Chicago, Illinois

INTRODUCTION

When the office of Keck and Keck began to pursue the development of the principle of using solar energy to help heat buildings, we found that there is nothing new under the sun. This is shown in the following quotation from Xenophon's Memorabilia, Book III, Chapter VIII, as translated from the Greek by Nino Repetto:

Again when he (Socrates) said about houses that beauty and utility were the same, he was giving a lesson in the art of building houses as they ought to be. He approached the problem thus: "Is it not true that to have the right sort of house it should be both useful and pleasant to live in?"

And this being admitted, "Is it pleasant," he asked, "to have it cool in summer and warm in winter?"

And when they agreed with this also, "Now in houses with a southern exposure, the sun's rays penetrate the porticoes in winter, but in summer, being less inclined, they afford us shade. If, then, this is the best arrangement, we should build the south side loftier to get the winter sun, and the north side lower to keep out the cold winds."

Not only were the Greeks aware of the principle, but the Chinese as long ago as 3000-3500 B.C., oriented whole villages to the south to help keep warm in the winter time. These prehistoric Chinese practices have been researched by Ken Butti and are discussed in the book *A Golden Thread*, by Butti and Perlin, published by Van Nostrand Reinhold in 1980.

Over a century and a quarter ago, an octagon house was built at Watertown, in the southeastern section of Wisconsin, not specifically as a solar house, but as an example of solving the problem of making life easier in the settlement of that portion of the state. The house is now owned and operated by the Watertown Historical Society and is open to the public for tours during the tourist season, spring through fall.

In its design it solved a number of problems in the use of energy. If these solutions had been developed during the later period of the growth of a domestic architecture in this country, it would have carried the development a lot further in creating adequate energy-efficient housing.

THE OCTAGON HOUSE

The octagon house was designed and built in 1850-52 by a lawyer who wished to settle in the midwest, with a promise to his future wife of "the best house in that part of the country." He came from the east where undoubtedly he had seen this type of house. His analytical mind told him that the greatest amount of enclosed space with the least amount of external area would be a circle. He, therefore, approached this by choosing an octagonal shape, which was simpler to construct. The brick he intended to use came from Milwaukee, a distance of 45 miles, and had to be hauled over plank roads by ox cart, since the railroads did not reach Watertown until 1855 or '56. The circle also gave the least amount of external area for heat loss.

To provide running water in the days before electricity, pumps and other plumbing developments, the house was built with a flat roof to collect rain water which flowed into an ingenious reservoir on the third floor of the building. This tank was constructed of wood and fitted with wedges which could be driven tight in case of an extended drought when the dry wood would shrink and leak. An overflow brought the water to a masonry cistern in the basement adjacent to the kitchen. The overflow from the cistern was wasted down the side of the hill. A lead pipe plumbing system was installed to provide running water with a line from the reservoir down to a coil in the kitchen stove and/or the heating plant, running back up to outlets in the stair hall as well as to a zinc lined bath tub on the second floor, providing running warm water for bathing.

A central heating plant was installed in the basement under the circular stairway. This stair ran only from the first floor up to the cupola at the roof level. A brick enclosure formed the housing in the basement, within which was placed a cast iron burner capable of taking four foot long cordwood. Ducts were run in the hollow brick walls of the staircase up to all the principal rooms of the house. At the rear of the housing was an access door through which pans of water were placed to help humidify the heated air.

The centrally located circular stairway provided an excellent duct work system to ventilate the house in the summer time. One only had to open the windows of the cupola on the lee side and open the windows on the windward side to provide a natural vent duct acting by gravity to bring in cooler air. Especially at night, comfortable air movement was easily induced without any mechanical ventilation.

The porches and roofs on the first two floors (now removed) provided the necessary sun control in the summer to help keep the house cool. Shutters on the windows provided summer shading and helped to maintain a warmer house in the winter by acting as storm sash.

The house was well thought out to provide greater comfort, utilizing ingenuity and taking advantage of natural phenomena.

Photo 1. The House of Tomorrow, 1933 World's Fair. KECK

THE HOUSE OF TOMORROW

In 1933 the Keck office designed a building for the Chicago World's Fair known as the Century of Progress. Named the House of Tomorrow, it was designed to demonstrate the most advanced construction in techniques and mechanical equipment available. The house was twelve sided (duodecagonal), with a central circular stairway around a utility core. Work shop, garage, recreation room, utility room and airplane hanger were on the ground floor; living and bedroom areas on the first floor, with a conservatory (solar room) above.

While the house was under construction in February 1933 (before the heating plant was in operation), but with the house enclosed against the elements, George Fred Keck visited the building on a sunny winter day with the temperature hovering between 0 and 10 Fahrenheit. Upon entering the living floor, he found carpenters and other workmen laboring in their shirt sleeves. Surprising as this was, when you stop to think about it, it was known that any greenhouse operator would bank his boiler in the morning of a winter day when the sun was shining, turning it on again only when the sun was gone for the day.

Based on these observations, the office tried to find more information on solar energy. None was available, in addition to which, the mechanical engineers stated that it would be too costly to heat such a house and that it would not be comfortable to live in. It was not until 1940 that the office was able to find actual data on solar gain.

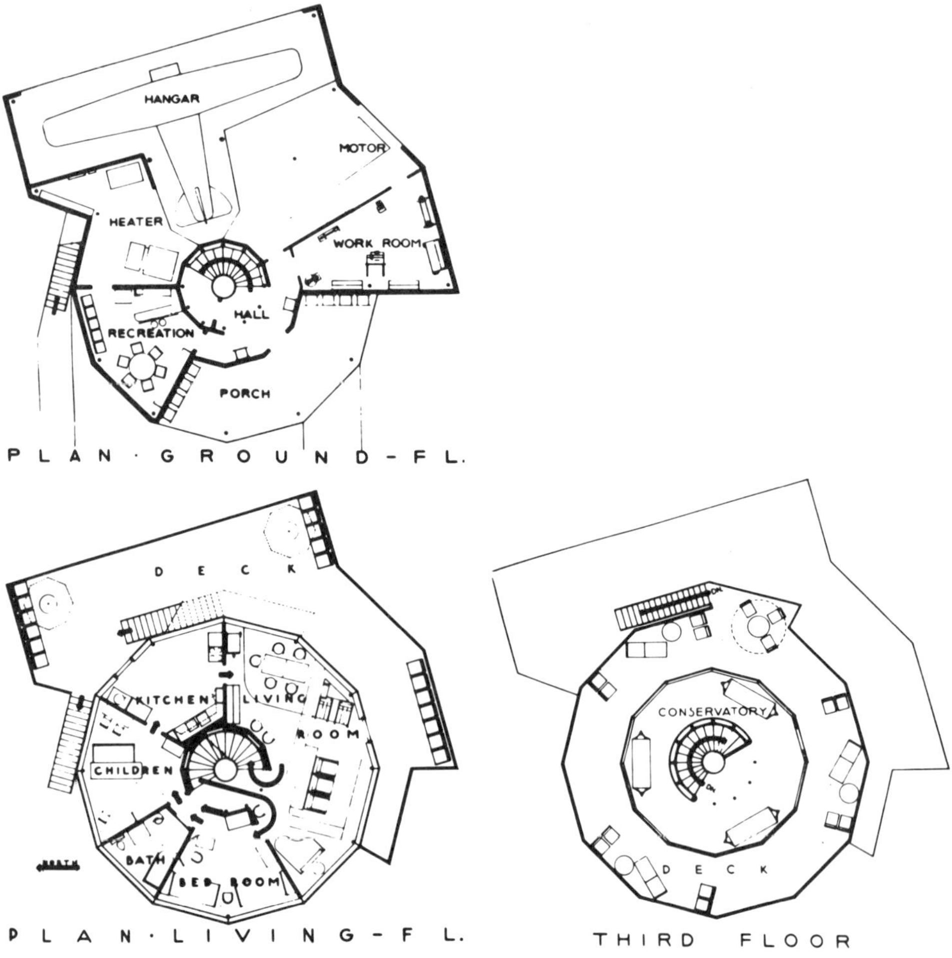

Figure 1. The House of Tomorrow, Floor Plan.

THE CRYSTAL HOUSE

The second year of the Century of Progress, 1934, a second house was built which consisted of an exposed prefabricated structural steel frame with glass walls around the outside on all three floors. It was called the Crystal House. No attempt was made, nor was it even considered in the design, to make it a solar house.

Photo 2. The Crystal House, Century of Progress, Chicago, Illinois, 1934

HEDRICH-BLESSING

EARLY HOUSES

In 1935 the office was commissioned to design a small house for a childless couple on a hillside lot on the north side of the Rock River in Watertown, Wisconsin. Because of the sloping lot and the location of the road, the garage and entry way, as well as the study or second bedroom, were located at street level, from which a half run of stairs led to the remainder of the house. The bedroom wing, therefore, was tucked into the hillside about half way, thereby reducing the heat loss to the north. Large windows were provided on the south side of the building. At this stage of the development of solar houses, we had not considered protection of the windows in the summer, using only nature in the form of trees to shade the windows.

The following year (1936) a house was designed for a professor at Whitewater, Wisconsin, Normal School (now one of the University of Wisconsin campuses). It was not specifically designed as a solar house; however, large windows to the south as well as to the east and west were protected by the use of metal venetian blinds on the outside of the house. The blind was so manufactured and installed that it operated in a track, could be tilted as well as raised and lowered from the inside, and was housed in a pocket above the sash if desired. This gave complete control of light as well as the elements and also provided some security against invasion of privacy or vandalism. This window treatment was the beginning of the development of some control against penetration of unwanted heat in the summer, spring and fall on large south-facing windows.

Photo 3. William H. Fricker House, Whitewater, Wisconsin. View from the southwest.

WILLIAM KECK

CAHN HOUSE

The B. J. Cahn House in Lake Forest, Illinois, was built in 1937. The Cahns at that time had an apartment in Chicago and wanted, as Mrs. Cahn stated to George Fred Keck, "not the House of Tomorrow, but the 'House of the Day After Tomorrow'." It was to be used as a week end house and/or, "on an hour's notice," one that could be opened at any time. The house was to have minimum maintenance; therefore, no curtains, as well as simple materials on the exterior, such as stucco, glass, lead coated copper and exterior venetian blinds of aluminum were used. Since Mrs. Cahn had a hip injury, walking with a limp, one of the requirements was no rugs or carpets on the floor to stumble over. This posed the problem of acoustics—glass, plaster, hard floors with the resultant use of acoustic ceilings. Much of the furniture, especially the dining room tables, were laminated with a sheet of aluminum to prevent cigaret burns. The wide overhangs to the curving form of the house provided excellent control of sun and light, and the use of external venetian blinds also provided control as well. Jen Jensen had done the original landscaping work for the Kuppenheimer estate on which this house was built which dictated a natural setting.

Photo 4. B.L. Johnson House, Columbia, Missouri. View from the south. WILLIAM KECK

JOHNSON HOUSE

In Columbia, Missouri, in 1938, a house was designed for B. Lamar Johnson on a sloping, wooded site facing a ravine. Again a sweeping plan with the main rooms of the house facing south. Since the site was heavily wooded, no overhangs were provided as protection. One innovation which was provided was an inch of water on the roof. Roof drains were formed with an adaptor ring which was removable so the water would drain away in the winter. A spigot with splash pan was provided above the roof line with controls at ground level to add water should normal rainfall be insufficient. The resulting evaporation helped considerably in keeping the house more comfortable.

Since large single panes of glass were used (double glazing such as Thermopane and Twindow not being on the market at that time), a

Photo 5. B.L. Johnson House. Showing roof cooling pond and kitchen clearstory windows.

WILLIAM KECK

continuous slot as a return grille under the windows was used to keep the cold draft from spilling out onto the floor. This had been done previously in a successful manner in the House of Tomorrow.

KELLETT HOUSE

In 1939-40 a house was designed for William Kellett in Menasha, Wisconsin, on a heavily wooded lot about four or five feet above the high water level of Lake Winnebago. No basement was provided because of the high water table. This was one of the first floor radiant heating systems used on the ground level to make the slab in the dining room, kitchen and maid's quarters habitable. In the large two story solarium it further acted as a "heat sink" to absorb and to store the sun's energy.

Thermopane and Twindow were still not in the market so we developed a system of windows for the house which had storage pockets above for the screen, the storm window and the inner sash. The large two story solarium windows, each 17'-6" high and about five feet wide, of course, were fixed. The large overhangs to protect these windows had pressure relief openings in them to prevent the roof from being blown off in a high wind.

The living room was on the second floor directly above the dining room and could be likened to a shelf within a box, with the open stair on the north side and the two story solarium to the south. This nine or ten foot elevation above the dining room gave a better view from the living room down the full length of Lake Winnebago to the south.

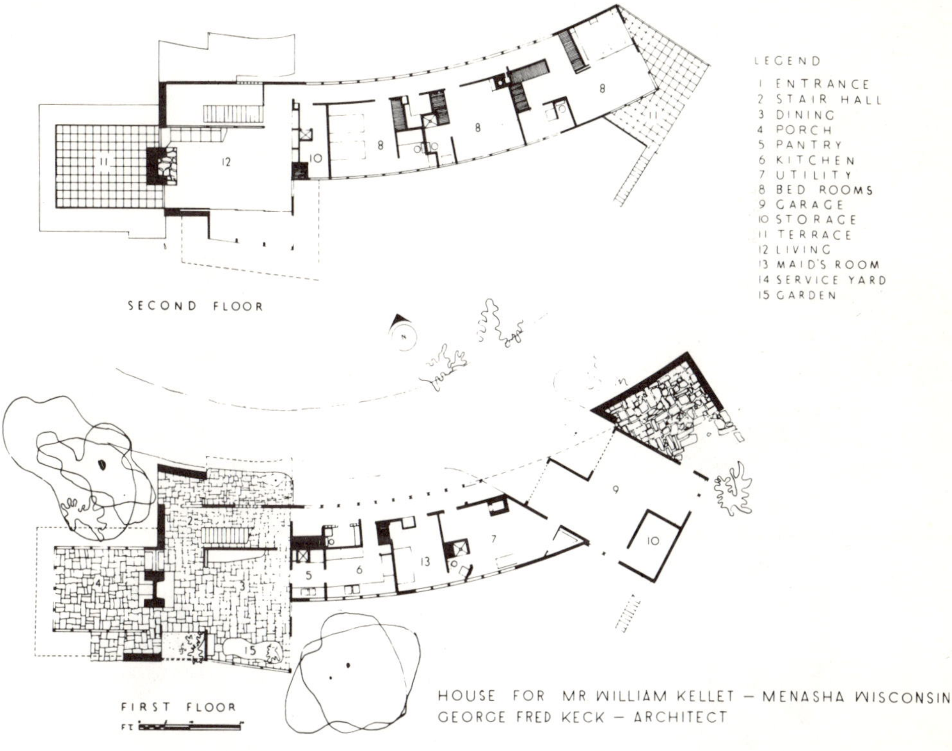

Figure 2. William Kellet House, Menasha, Wisconsin. Floor Plan.

In 1940, while attempting to find more information on Solar Energy, the office finally wrote to the U. S. Weather Bureau in Washington, D. C. They informed us that the University of Chicago was operating a weather station in Chicago, tabulating solar energy gain behind glass throughout the day. A visit to their facilities showed that they were using what appeared to be about the size of a 200 watt light bulb, with the stem down exposing the globe to the sun. On the inside was a small black area connected with electrodes to a recording device which gave readings over the day of gram calories per sq. centimeter. Taking the readings for five sunshiny days in February 1940, we came up with the following average tabulation:

Photo 6. William Kellet House, Solar Room. HEDRICH-BLESSING

Time	9–10a.m.	10–11 a.m.	11–12 noon	12–1 p.m.	1–2 p.m.	2–3 p.m.
Gr.Cal./Sq.Cm.	13.96	20.2	24.7	27.0	23.6	16.8

SLOAN HOUSE

A developer and contractor on the North Shore by the name of Howard M. Sloan built a solar house in 1940 in Glenview, Illinois. Although this house was initiated without being called a solar house, Thermopane and/or Twindow were now back on the market, after having been withdrawn for a while because of technical failure. Thus, here large areas of double glazing were used. But large areas of double glazing with their sealed edges were subject to rupturing whenever and if stress were placed on the glass. So the office decided to divorce the functions of light and ventilation, fixing the glass in place and providing ventilation through louvers, behind which on the inside was a screen and a weatherstripped wood door. This gave better control of ventilation as well as security from prowlers in a one story house. Sloan eventually built quite a number of this type of house in the area. In promoting the houses, the Real Estate Editor of the Chicago Tribune dubbed this type of house a "solar house." This was the first time, to the best of our knowledge, that the phrase was used in connection with a house exposed to a large amount of sunshine.

Photo 7. Sloan House, Glenview, Illinois. The first home to be dubbed "Solar House"

HEDRICH—BLESSING

PULLIAM HOUSE

The following year, 1941, a house was designed for the Pulliams in Lake Forest, Illinois. Again, it was a small one story house with the principal rooms facing south. Subsequently, the house was made larger with an additional two bedrooms, and recently, the outside screened porch with its own barbeque fireplace, was enclosed and enlarged as a Solar Room. The house had only single plate glass with smaller type vent sash. Except for the fact that some condensation does occur if the humidity is relatively high in colder weather, the building had a good record on low cost for energy for heating. The house had a concrete slab on the ground with an air system of heating through flue tile under the slab, giving a warm floor plus the mass for absorption of solar heat as a storage means.

In the same year, 1941, a variable level house was designed for the John Bennetts near Barrington, Illinois, on a 40 acre site. Using a local glacial deposit for stone work in combination with painted vertical siding, an interesting effect of tying the house into the hillside was obtained. Part of the living area was a story and half in height, giving a feeling of spaciousness. The floors of the living and bedroom areas were finished with black slate set over radiant heating coils in a concrete slab. Again, large areas of double-sealed glass were used on the south elevation.

DUNCAN HOUSE

The Duncan house in Flossmoor, Illinois, was built in 1941 for a young sociologist professor and his wife. The original plan (which was later enlarged) had one bedroom, a combination study, living and dining room, along with a kitchen facing south, protected by a corridor and storage-utility room to the north. Radiant hot water floor heating was installed in a concrete slab on the ground, which was stained a dark brown to absorb heat. Double glazing was used in the sash. Libbey-Owens-Ford Glass Co. became interested in sponsoring a year's test to be operated by the Illinois Institute of Technology. Results were published in the *Architectural Forum* issue of August, 1943. It was concluded that, using the same amount of glass oriented in a southerly direction, a savings of from 15 to 20 per cent could be anticipated. Overhangs to the south were provided for sun protection. In addition, some vertical adjustable wood louvers were provided at intervals on the south elevation to control early morning and late afternoon sunshine at certain times of the year. These were adjustable from the inside.

Again in 1941, a house was designed for Dr. Emil Quenneville in Granby, Province of Quebec, Canada, on a rocky site, using brick and wood, with a high ceiling living room.

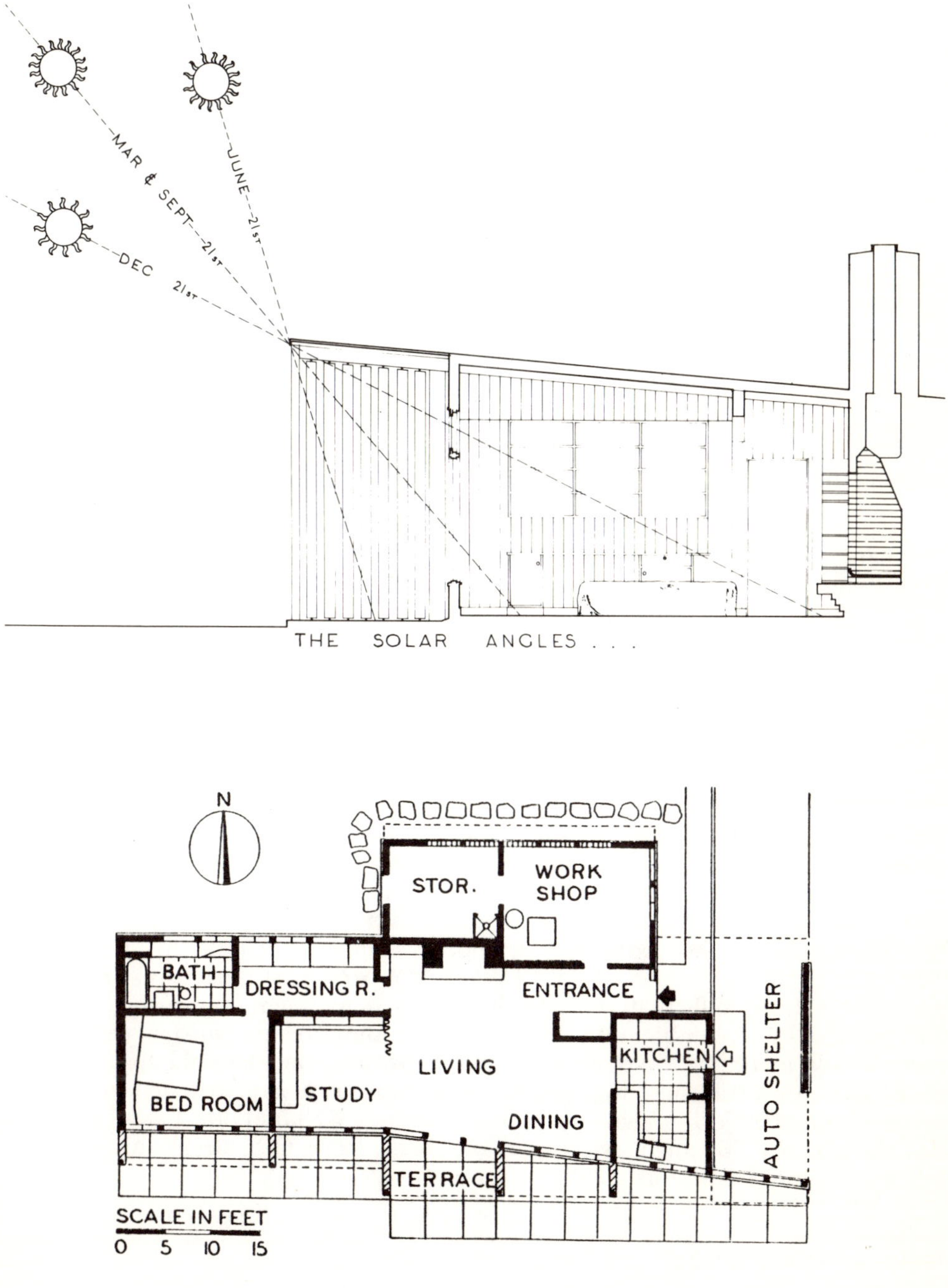

Figure 3. The Duncan House, Flossmoor, Illinois. Plan and Elevation.

WORLD WAR II

The onset of World War II in December 1941 brought a halt to much building except under priority. During the early stages of the war while William Keck was in service, George Fred Keck approached the Green Company in Rockford, Illinois (builders of prefab panels for war housing and shell boxes) on the possibility of developing a prefabricated method of construction for houses rather than large prefabricated panels assembled on the site. The outcome was the granting of a priority by the government to design a method and to build a model of a prefabricated sectional house in Rockford, Illinois. Working with the Forest Products Laboratory in Madison, Wisconsin, the Green Prefabricated House was developed. Thirty-nine inch wide panels (approximately 1 meter) were used for the interchangeable wall sections, 8'-0" high. Roof sections were the same width, twelve feet long. Electronic glue welding of plywood, studs and/or roof joists increased the structural strength to provide a prefinished panel capable of being bolted together on the site to form a rigid structural unit.

In addition, a floor radiant heating system was developed with the Clay Products Association, using warm air in a closed system under the floors, with hollow floor tile to provide uniform floor temperatures. Metal was scarce during the war, with ample supplies of clay which led to the development of this method of heating.

THE WHITEHORN HOUSE

The Whitehorn house, built in a suburb of Detroit, Michigan, in 1949, posed an interesting problem since the owners had written to a number of modern architects, including Lescaze, Neutra and F. L. Wright, along with this office, asking if they were interested in building a $10,000.00 home, and outlining their requirements. Wright returned their letter with a message scrawled across the bottom, stating: "There is no such thing as a $10,000.00 house." Signed, "F.L.W." We wrote stating that it was a formidable task, but that we'd be happy to help them. The plan included a car port with a combination living-dining room opening into a den at one end, which could be used as a second bedroom; one large bedroom with a split bath for efficiency was provided, as well as a large kitchen and utility room, and the job ended up costing about $13,500.00. Large panes of double glazing, 96" high, were used on the entire south elevation except for some louvered ventilators.

THE LEVIN HOUSE

The Levin house in Champaign, Illinois, built in 1949, is a large two story house, again a solar example. The University of Illinois is now making an analysis of the house on heat loss, heat gain, etc. It will be interesting to see the result. We can only quote from an unsolicited letter written by Mrs. Helen Levin in 1976 in which she states: "It is 27 years since you planned our house. The design is a constant pleasure; its function satisfies our changing life styles. The architect is the non-present person who is an integral part of family living in a home. And we thank you again for the house. You will be interested to know that *after all these years*, ours will be included in a 'Contemporary Open House'—a project of opening homes to the public as examples of contemporary architecture."

MULTIPLE HOUSING

An application of passive solar heating is shown in an apartment building known as Pioneer Co-op on the south side of the City of Chicago, built in 1950. Here in a four story building the office produced a plan with south balconies on each of the living areas of four apartments per floor. Using large amounts of glass, the overhangs protect the rooms in the summer, yet allow penetration of the sun in the winter.

The same application was used on a Chicago Housing Authority project known as Prairie Courts, using a south balcony access corridor to the apartments in the fourteen story building. Two sets of elevators were used in order that the traffic in front of the living areas of each of the apartments was minimized. The balcony provided ample sunshine in the winter yet shading the living areas in the summer. It further provided sitting space outside on the balcony for high rise tenants.

GROSSMAN HOUSE

The Grossman house, built in 1950 in North Muskegon, Michigan, was located on the north side of Lake Muskegon, a smaller lake off Lake Michigan, thus providing an ideal site for a solar house. All the principal rooms face south, including the bedrooms. The children's bedrooms were below the master suite and guest room, part way into the ground, thus saving energy on heating. Hot water radiant heating was provided in both the ceiling and floor of the living room because of the high ceiling. This provided the necessary mass of concrete to absorb some of the sun's rays. Instead of providing a solid overhang to protect the windows in the summer, a system of wood louvers was used which allowed light to penetrate, yet shielded the direct rays of the sun from the window panes. Because of the physical location of the house on the east side of Lake Michigan, where the prevailing winds are from the southwest in the summertime, little or no air conditioning is needed because of the passive effect of the winds being cooled as they pass over the Lake.

KUNSTADTER HOUSE

The Kunstadter house, 1951, in Highland Park, Illinois, is situated on a ravine just off Lake Michigan. Again a concrete slab on the ground, heated by hot water radiant heat, provided the mass for absorption of the solar energy. A simple rectangular form, relieved slightly by varying depths of overhang projection, with a detached garage, gave most of the rooms a southern exposure. The house was mainly of wood and glass with only a limestone chimney for the fireplace. The house won an Honor Award from the American Institute of Architects in 1953.

BLAIR HOUSE

The Blair house in Lake Bluff (1953) was on the bluff above Lake Michigan on a large wooded piece of property about sixty feet above the lake. Since the view was of great importance, the main rooms face east. However, a large solarium was provided at the south end of the building, two stories high, which opened to the living room on the ground floor and the master bedroom on the upper level. Glass sliding doors were used at the upper level to control both sound and the warmth of the sun's rays into this room. Water ponding was used on the roof to help cool the house without air-conditioning.

BORTIN HOUSE

In Mequon, Wisconsin, a suburb of Milwaukee, a house was designed for the Bortins (1958). Using concrete block as a simple, less expensive material, along with glass on the south side, the office created an excellent solution for allowing the sun to penetrate the house. Again, a concrete slab was used on the ground with forced warm air in a system of duct work which helped keep the floor warm and acted as a "heat sink." An unsolicited letter from Mrs. Bortin in 1976 read, in part: "Dear Mr. Keck:----You'll probably be interested in the fact that we received all our heat from the sun during daylight yesterday, when the outside temperature was 22 below zero."

BLOOM HOUSE

The Bloom house (1959) is in a similar location to the Grossman house, mentioned above, in North Muskegon. Because of the location rather close to the lake level which varies over periods of years, the house was designed as though it was in a flood plain. Only an entrance with stairway and the heating equipment are on the ground floor. The entire living area is up one flight, including a screen porch. The floor was well insulated with hot water heating coils in a concrete slab poured over the wood joists. The entire south area was restricted somewhat in the amount of glass, with sufficient amount, however, to provide passive solar radiation.

ISAACSON HOUSE

In 1959 the Isaacson house was built as an experimental prefabricated house of steel. Steel plates 5/16" thick, 4'-0" on center, with edges bent up to give stiffness, formed the slightly sloping roof as well as the side walls. Where the upturned edges met on the roof a U-shaped piece was welded to the flanges to make them water tight. Holes were punched in the panels to the house and for doors elsewhere, with wood frames inserted to provide support for glass and/or louvers. The entire house was erected on a pre-installed foundation within three days' time. The interior of the ceiling panels were sprayed with insulation to form an acoustical ceiling. The walls were also sprayed and then finished with plaster or other materials.

One must remember that not until the energy crisis in the 1970's (when the price of fuel escalated considerably) it was uneconomical to provide more than 3 or 4 inches of insulation in the walls and ceilings. Beyond that point the savings in fuel on your additional investment would not pay for the extra cost.

A recent house (1977) for the Wolfs in Munster, Indiana, provided good solar orientation in the living room, den and master bedroom area, using a louver system of overhangs similar to the Grossman house in North Muskegon.

In 1978 a Solar Room or Green House was added to the row house for the Pickens in Chicago. Constructed of double glass for the side walls and three plastic double sky domes for the ceiling, the room is a pleasant addition to the dining room of the house. Unfortunately, it is not a true Solar Room because it is on the north side of a series of row houses. The Pickens, however, have an excellent collection of African violets and other plants which need light more than sunlight.

The development of the use of passive solar energy by the office was, as you can see, accomplished over a period of years. No great attempts were made to store the energy other than to trap it within the building and absorb it in the mass of the floor construction. Control became important since many clients would call the office after they were in the house to say that the heating plant was off and that it was overly warm in the room. The usual answer was "open a window." As was stated earlier, energy was inexpensive! The effects other than solar gain were important in that better illumination in the area exposed to sunlight had some therapeutic effect on better eyesight and in creating a feeling of well being and relaxation.

The increase in costs of energy have now made it a necessity to find ways and means of conserving energy as well as to find alternate sources. This becomes the job of the younger generation who will have to solve these problems.

In closing, we would like to give you a quotation from the book, *Man and Sun*, by Jacquetta Hawkes, in which she was speaking about the expansion of use of solar energy in recent years, stating as follows: "This direct utilization of the sun's radiations will increase from year to year—though it is never likely to equal that quietly carried on by the vegetable world."

KECK AND KECK BIBLIOGRAPHY

Based on a bibliography initiated by Susan Sten, student in architecture, Rice Institute, Houston, Texas, 1979.

1. Cohen, Stuart E., *Chicago Architects*, Chicago: Swallow Press, Inc., 1976.

2. Daniels, George, *Solar Homes and Sun Heating*, New York: Harper and Row, 1976.

3. Ford, Katherine M. and Thomas H. Creighton. *The American House Today*. New York: Reinhold Publishing Co., 1951.

4. Ford, Katherine M. and Thomas H. Creighton. *Quality Budget Houses*. New York: Reinhold Publishing Co., 1954.

5. Hitchcock, Henry R. and Philip Johnson, *Modern Architecture: International Exhibition*, New York: for the Museum of Modern Art, by Arno Press, 1969 reprint edition.

6. Moholy-Nagy, Laszlo, *Vision in Motion*, Chicago: Paul Theobald, 1947.

7. Simon, Maron J., editor, *Your Solar House*, New York: Simon and Schuster, Inc., 1947.

8a. Watson, Donald, *Designing and Building a Solar House*, Charlotte, Vt., Garden Way Publishing Co., 1977.

8b. Wingler, Hans M., *The Bauhaus*, Cambridge, Mass.: M.I.T. Press, 1969.

9a. *American Architect*, July 1934; p. 40, "Trends and Topics."

9b. *Architectural Digest*, January/February 1973; p. 32-39, "A House For All Seasons", Walter Placko House, 1955.

9c. Menocal, N. and Narciso, G. *Keck and Keck Architects*. Madison: Elvehjem Museum of Art, University of Wisconsin 53706. $10.

Architectural Forum:

10. V.60; February, 1934; (Supplement), "House of Tomorrow."

11. V. 71; July, 1939; p. 12-15.

12. V. 71; October, 1939; p. 262.

13. V. 72; April, 1940; pp. 248-9, "House in Ogden Dunes, Indiana."

14. V. 74; June, 1941; pp. 412-13, "Four Bedrooms...for Summer Use." (House in Northport, Michigan)

15. V. 75; October, 1941; pp. 54-7 (Supplement).

16. V. 77; August, 1942; pp. 67-82, "A Portfolio of Modern Houses", (Houses in Flossmoor, La Grange, & Highland Park, Illinois; Columbia, Missouri; and Menasha, Wisconsin.)

17a. V. 77; September, 1942; pp. 138-9, "Housing Standards."

17b. V. 78; August, 1943; pp. 6, 7, 8, "Solar Heating—Survey Proves Large Windows, Properly Oriented, Save Fuel Even in Rigorous Climates."

18. V. 80; March, 1944; pp. 85-94, "Two Developments." (Residential developments of Meadowbrook and Solar Park, in Glenview, Ill.)

19. V. 82; June, 1945; pp. 98-9, "House in Steven's Point, Wisconsin."

20. V. 86; April 1947; pp. 102-3, "Lake Shore House..." (Pete Keck House)

21. V. 87; September, 1947; pp. 107-9, "This Midwest House..." (Pete Keck House)

22. V. 88; May, 1948; p. 133.

23. V. 95; October, 1951; p. 252.

24. V. 96; January, 1952, "The Case for the High Apartment." (Prairie Avenue Courts.)

Architectural Record:

25. V. 67; February, 1930; pp. 105-9, "Miralago Ballroom and Shops."

26. V. 71; March, 1932; pp. 167-208.

27. V. 75; January, 1934; p. 29, "House of Tomorrow, Century Homes, Inc., Chicago, George Fred Keck, Architect."

28. V. 77; March, 1935; p. 20l, "House at Kaukauna, Wisconsin."

29. V. 79; April, 1936; pp. 314-17, "House of E. L. Wilde." (Watertown, Wisconsin.)

30. V. 83; February, 1938; pp. 31-7, "Architect Stages 1-Man Design Job in Chicago House." (Herbert Bruning house in Wilmette, Ill.)

31. V. 87; May, 1940; pp. 41-44, "New Apartments Stress Flexible Planning." (Architect's own residences at 5551 S. University, Chicago.)

32. V. 89; January, 1941; pp. 91-93, "Houses." (A "Solar House" for Howard M. Sloan, Glenview, Illinois.)

33. V. 92; November, 1942; pp. 65-8.

34. V. 92; December, 1942; p. 40.

35. V. 95; March, 1944; pp. 58-63, "Solar House for a Sunny Hilltop." (Residence for M/M John L. Bennett, Barrington, Illinois.)

36. V. 96; December, 1944; pp. 70-1, "Three Houses for the Postwar World."

37. V. 101; May, 1947; pp. 132-3, "Straight Lines or Panoramic Curves." (Houses for Dr./Mrs. Irving Cowan, Milwaukee, Wisconsin; and for M/M Jerrold T. Kelly, Barrington, Illinois.)

38. V. 104; September, 1948; pp. 92-6, "Broad in Vision Though Narrow in Land." (Residence on the shore of Lake Michigan.)

39. V. 109; March, 1951; pp. 105-10, "A House with 'Emotional Content' " (House for M/M Abel E. Fagen, Lake Forest, Illinois.)

40. V. 110; December, 1951; p. 146, "Chicago Cooperative has Exterior Corridors." (Cooperative Residences, Inc., Chicago, Ill.)

41. V.113. January, 1953; p.134-6. "Designed for Informal Living." (Home for William E. Neuman, Du Page County, Ill.)

42. V. 115; February, 1954; pp. 196-201. "Zonate Plan Yields Amenity in Living." (Sigmund Kunstadter House in Highland Park, Ill.)

43. V. 115; May, 1954; pp. 150-3, "An Innovation in Open Planning." (Residence for M/M John C. Telander, Hinsdale, Illinois.)

44. V. 115; June, 1954; pp. 176-8, "Chicago: Wide Variety of Apartment Sizes." (Pioneer Cooperative, Inc.)

45. V. 116; October, 1954; pp. 170-7.

46. V. 117; March, 1955; p. 176, "Chicago Cooperative Supermarket."

47. V. 118; November, 1955; pp. 168-71, "Michigan House Puts Bedrooms on Lower Level." (House for M/M Robert Feldman, Benton Harbor, Michigan.)

48. V. 119; January, 1956; pp. 170-97.

49. V. 120; November, 1956; pp. 161-86, "Interior Court, Terraces at Rear." (M/M Herman Grossman residence, North Muskegon, Mich.)

50. V. 121; February, 1957; pp. 183-88, "Residence of Exceptional Distinction." (E. McC. Blair House, Lake Bluff, Illinois.)

51. V. 123; May, 1958; pp. 122-7. (Walter D. Gray Residence, Olympia Fields, Illinois.)

52. V. 124; October, 1958; pp. 175-6, "Small Medical Clinic on a Sloping Site." (for Dr. Robert E. Reagan, Benton Harbor, Mich.)

53. V. 126; December, 1959; pp. 125-8.

54. V. 127; June, 1960; pp. 174-6, "Economical Design for Research." (Laboratory for National Clay Pipe Manufacturers, Inc., Crystal Lake, Illinois.)

55. V. 129; February, 1961; pp. 133-6, "An Enclosed Court Highlights a New Keck House." (Residence in Kenilworth, Illinois.)

56. V. 130; November, 1961; pp. 156-8, "A Raised Cottage for a Lakeside Site." (Dr. Robert E. Bloom Residence in North Muskegon, Mich.)

57. V. 131; May, 1962; pp. 102-5, "Curved House Designed for Sweeping View." (Payne Residence in Pleasant Valley, Pa.)

58. V. 133; May 1963; pp. 98-101. (Milton Hirsch House in Highland Park, Illinois.)

59. V. 137; February, 1965; pp. 157-60, "Sliding Roof Gives House Year Round Use of Patio." (M/M Norman Weinrib Residence in Highland Park, Illinois.)

60. V. 137; April, 1965; pp. 106-7, "Buyers Prefer Four-Bedroom Town Houses." (Townhouses, 48th & Woodlawn Ave., Chicago, Ill.)

61. V. 138; November, 1965; pp. 159-60, "Concrete for Shade and Pattern." (Design for Child Care Center, Chicago.)

62. V. 139; May, 1966; pp. 64-7, "Schramm Residence, Burlington, Iowa."

63. V. 141; February, 1967; pp. 143-6, "A House Planned for Cars as Well As People." (Teplinsky Residence, Highland Park, Ill.)

64. V. 141; May, 1967; pp. 76-7. "Karlin House, Chicago."

Architectural Review:

65. V. 81; March, 1937; pp. 116-17.

66. V. 88; December, 1940; pp. 175-8.

66a. V. 158; November 1973; pp. 332. "Chicago Child Care Society & Harper Square Housing Project".

67. V. 162; October, 1977; p. 256, "Dr. Robert A. Wolf House, Munster, Indiana."

67a. *Architecture + Urbanism (Tokyo).* September 1977. "House of Tomorrow, Herbert Bruning House and B. J. Cahn House".

67b. *Art in America,* March 1976; "Report from Chicago-Architecture, City & Town Views".

67c. *Chicago Daily News.* 6/7 March 1976. "Chicago Architects: The New Show in New York."

67d. *Chicago Magazine.* April 1976. "The Makers of the Chicago School—Keck and Keck." by Don Kilmovich.

67e. *Chicago Sun Times,* 17 February, 1974. "Let the Sun Keep You Warm".

67f. *Chicago Tribune,* 7 April 1968. "Architect Specializes in Design of Homes". Micro S530.

68. *Domus.* No. 278; Gennaio, 1953; p. 20.

69. *House Beautiful.* V. 84; July-August, 1942; pp. 30, 72.

70. *House and Garden.* V. 86; November, 1944; pp. 80-l.

71. *House and Home.* V. 14; July, 1958; p. 104, d & e.

72. *Inland Architects.* July, 1965; pp. 9-24, "George Fred Keck, William Keck: Architects."

73. *Interiors.* V. 100; March, 1941; pp. 32-5, and 61-2.

74. *Interiors.* V. 105; May, 1946; pp. 82-93.

75a. *Journal of American Institute of Architects.* V.33; June 1960, pp. 65-6.

75b. *Journal of American Insitute of Architects.* V.68; No.14, Dec., 1979; pp. 48-9.

76. *Journal of Society of Architectural Historians.* V. 29; No. 4; December, 1970; p. 350-3, "The Crystal House of 1934."

76a. *Los Angeles Times,* 26 April, 1977. "A Window on Chicago Architects". Micro S1058.

76a. *Los Angeles Times,* 26 April, 1977. "A Window on Chicago Architects". Micro S1058

76b. *New Art Examiner,* May 1976. "New York Looks at Chicago Architecture". by Carol Diehl.

76c. *New Art Examiner,* June 1976. "City of Towers Exposed—Two Architecture Shows."

Pencil Points:

77. V. 24; May, 1943; pp. 44-9. (M/M Richard E. Pulliam House, Lake Forest, Illinois.)

78. V. 25; February, 1944; pp. 76-82, "Insolation and House Design." (Howard M. Sloan House in Glenview, Illinois.)

79. V. 25; December, 1944; p. 112, "Review: The Solar House by G. F. Keck."

80. V. 26; September, 1945; pp. 60-6, "Houses for the People." (Houses designed for the *Ladies Home Journal.*)

81. V. 27; July, 1946; p. 64-7, "House in Oconomowoc, Wisconsin." (Pete Keck Residence.)

82. V. 32; April 1951; pp. 57-67, "Prairie Avenue Courts."

83. *Progressive Architecture*, Feb. 1981."George Fred Keck, 1895-1980", p.22.

84. *Solar Age*, March 1981.

85. *Vitrum*, Milan, July/August 1967, "Child Care Society, Chicago".

Historical Applications of Solar Architecture

Ken Butti

This article is abridged from A Golden Thread: 2500 Years of Solar Architecture and Technology, *Van Nostrand Reinhold Co., 1980, New York.*

INTRODUCTION

The current energy crisis has increased interest in passive solar architecture. As a result, over the last several years more and more information has been published regarding passive solar design. Yet, relatively little has been written specifically about solar fenestration, a vital consideration in solar design, and little has been written about the ancient and widespread historical use of such design. The purpose of this paper is to briefly trace the evolution and past use of such architecture.

GREEK SOLAR DESIGN

One of the earliest written accounts of solar architecture comes from ancient Greece.[1] The Greeks did not have any mechanisms for cooling their homes in the extreme heat of summer, nor did they have an adequate heating system for the cold winter. They frequently used portable braziers fueled by charcoal for heat.

By the fifth century B.C., wood and therefore charcoal had become very scarce. The Greeks, having consumed much of their domestic supply, depended more and more on costly imports with supply lines stretching all the way across Asia Minor to Russia.[2]

During this time, it became popular to utilize the sun's energy for heating buildings. In many parts of Greece, the use of the sun was a direct response to the fuel shortage. Living in a climate that was sunny almost year-round, the Greeks learned to build their houses to take advantage of the sun's heat during moderately cool winters, and to avoid its heat during the hot summers.

Solar architecture flourished throughout ancient Greece, as modern excavations indicate. Not only were individual homes oriented south, but entire cities were also planned to allow their citizens equal access to the winter sun. A solar-oriented home allowed its inhabitants to depend less on charcoal-burning stoves—conserving fuel and saving money. Thus solar architecture was economically appealing.

Olynthus, one of the leading cities of northern Greece during Hellenic times, illustrates how the Greeks put solar architecture into practice. The major portion of Olynthus was totally planned, as were most Greek urban areas at the time. Situated atop a sweeping plateau, the streets of Olynthus were laid out perpendicular to one another, running north-south and east-west (Figure 1). In this way, all the houses on a street had a southern exposure so they could take maximum advantage of the sun.

The Greeks knew that in winter the sun travels in a low arc across the southern sky and in summer, it passes high overhead. They built their homes accordingly. The main rooms of the Olynthian house faced south onto a portico and caught winter sunlight through small window openings and doorways. The north side of the home was sheltered against cold winter winds and had few if any windows. Eaves on the south side of the home shaded rooms from the high summer sun through most of the day.

In winter, rays from the sun streamed across the south-facing portico, in through the home's window openings and doorways, heating the main rooms. The earthen floors and adobe walls absorbed and retained much of this solar heat. In the evening, when the home began to cool down, stored solar heat within the floors and walls helped warm the house. For the majority of the winter, the sun would adequately heat the main rooms of the home during the daytime. On those occasions when solar heat was insufficient, charcoal stoves would be lit.

The Olynthian solar house also worked well in summer. The home's south-facing portico shaded the main rooms from the harsh heat of the high sun. In addition, closed adobe walls and other contiguous dwellings on the east and west sides protected the house from morning and afternoon sunlight.

Excavations throughout Greece indicate that solar architecture prevailed in many parts of ancient Greece. Solar design cut across class lines; rich and poor city dwellers, as well as princes and kings relied on the sun. All of the important ceremonial and state rooms of the Palace at Pergamon, for example, fronted south, just as the living rooms of commoners at Olynthus. Many leading archeologists familiar with classical Greece agree that solar architecture was a prime concern of fifth-century Greek builders.

ROMAN SOLAR DESIGN

Like Greece, Rome also depleted fuel resources at a voracious pace. For example, at a short distance from Rome, Monte Cimino was a densely forested region up to the third century B.C., but by the first century B.C., Rome had to import pine and other types of wood from as far east as Caucasia. Fuel consumption continued to rise with the standard of living. By the first century A.D., many of Rome's wealthier citizens enjoyed central heating. Such systems could consume up to 287 pounds of wood per day.[3]

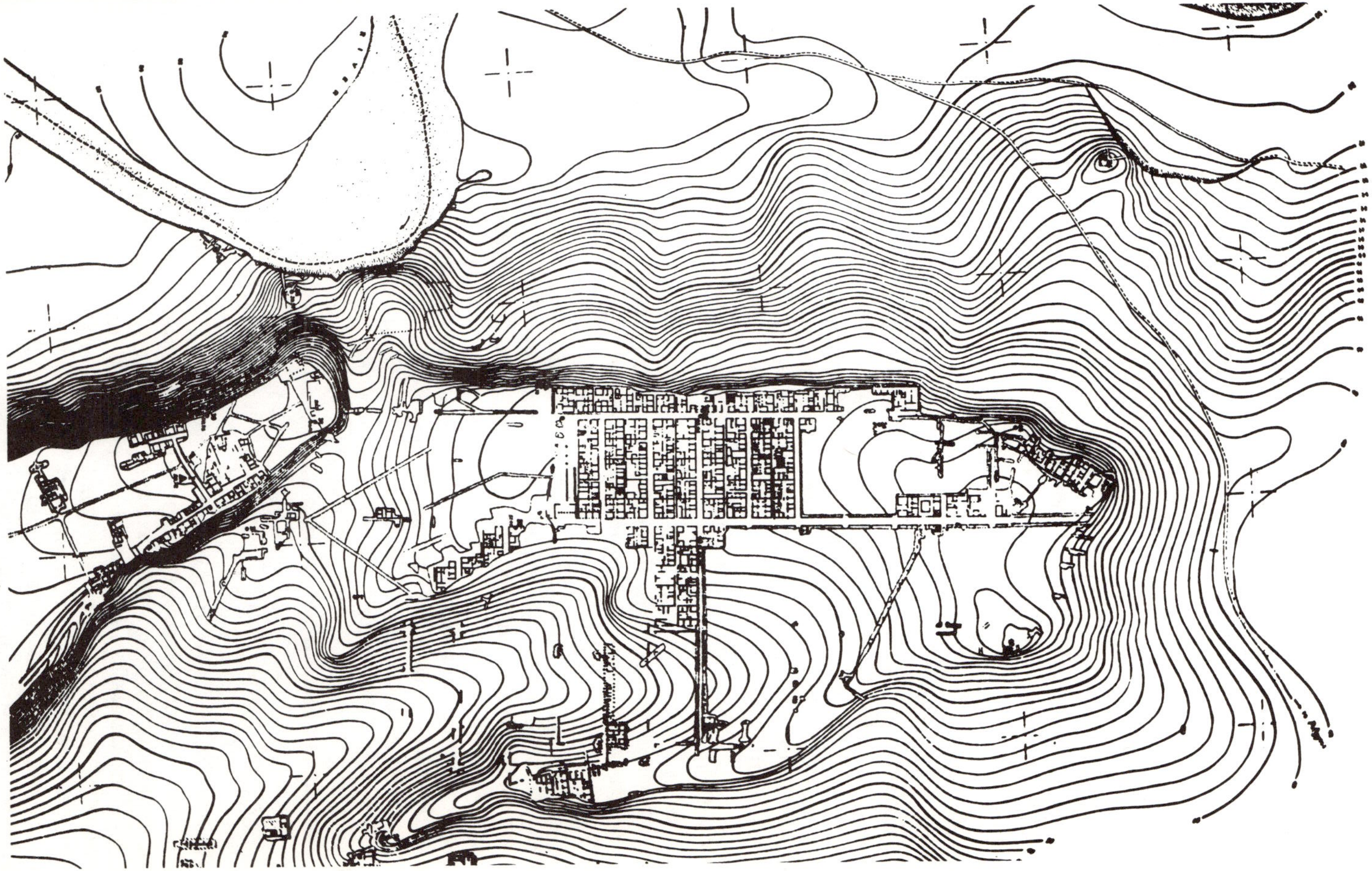

Figure 1. Circa 432 B.C., Olynthus, Greece. Illustration of Olynthus' orthogonal plan. All homes fronted south.

Local fuel shortages and the high cost of imported wood probably influenced the Romans to adopt the Greek techniques of solar architecture. They did not simply copy the Greeks, but brought solar technology several steps further. The Romans discovered that the use of clear windows—such as glass and mica—enhanced the effectiveness of solar heating. They also expanded solar architecture to include not only homes, but greenhouses and huge public bath houses as well. Solar architecture became so much a part of the Roman way of life that sun-rights laws were eventually enacted.

As timber supplies dwindled from the first century A.D. onward, wealthy Romans throughout the Empire began building with the sun in mind. For those with country villas, a solar-oriented home required a smaller furnace and consumed less fuel than a non-solar home. This was attractive to the Roman citizen worried about the rising cost of fuel and its eventual depletion.

Main windows in such homes usually faced south. But certain rooms oriented east or west of south to catch the winter sun at the time of day those rooms were in use. In the villa of Pliny the Younger, for example, a wealthy Roman writer, his study was semi-circular with large bay windows facing southwest to catch the winter sun in afternoon. Pliny called this room his *heliocaminus*—literally "solar furnace"[4] (Figure 2). The southwest openings of Pliny's *heliocaminus* were probably covered with transparent glass or mica in order to trap solar heat.

Wealthy home owners had little trouble purchasing window covering for their *heliocaminus*, since a thriving window industry existed in Rome at the time.[5] Transparent coverings made of thin stone such as mica or selenite were produced by splitting the stone into plates as thin as desired. Glass windows are thought to have been fabricated by pouring molten glass into molds or onto a flat surface and pressing it with a moist wooden mallet. Another process involved blowing hot glass into a bubble, swinging it until it formed a cylinder, and cutting it with glass shears before flattening it out on an iron plate covered with sand. The Romans could manufacture glass windows as large as 40 by 28 inches.

Another important application of the solar heat trap principle discovered by the Romans was the use of transparent window coverings to help heat their public baths. Previously baths had only tiny "chinks" cut out of stone to let in air and light. But in solar baths, at least in the hot bath or *caldarium*, large glazed windows faced southwest toward the winter sunset. In addition, the Romans usually glazed the whole south wall of their bath houses. These giant windows trapped so much solar heat that by the late afternoon bathers would, as one early writer put it, "broil" inside the baths. The use of solar design cut down considerably the tremendous amount of fuel needed to keep the bath water hot, the steam room at a high temperature, and the air inside the building comfortable.

Roman law also took into account the sun rights of its citizens. By the second century A.D., it had become a civil offense for anyone to place an object in such a way that it obstructed the solar exposure to windows which required access to the sun's energy.[6]

The mass scale use of the sun, however, never caught on in the Roman Empire as it had in Greece. Rome did not plan her cities to take account of her poorer citizens—the dominant ideology favored class privilege in contrast to the egalitarian spirit of Greece—so that only the upper class could afford to hire designers to build their homes with proper solar orientation.

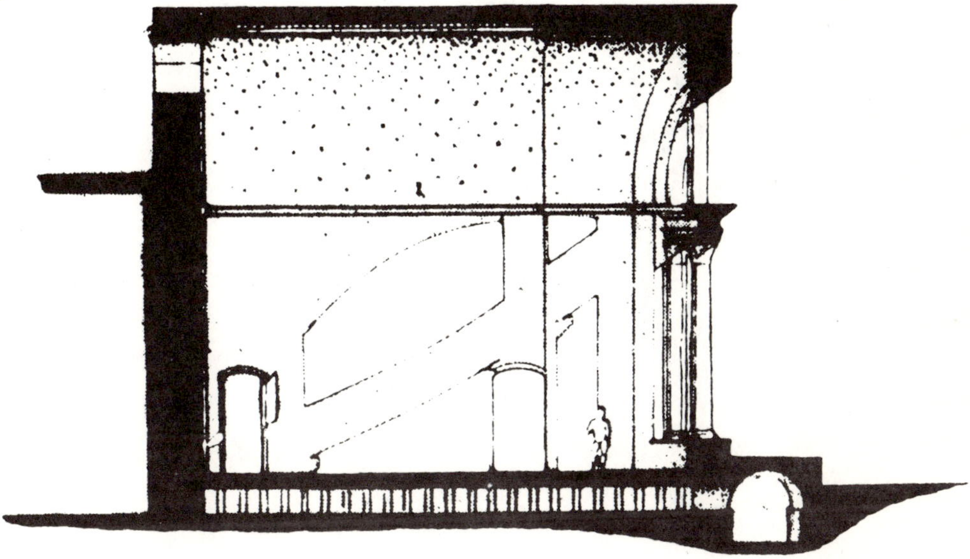

Figure 2. Ostia, Italy. Rendering of a heliocaminus similar to the one described by Pliny. Sunlight easily penetrated the large windows at the right and entered the sunning chamber (the white pattern on the wall represents the sunlight).

CHINESE AND JAPANESE SOLAR DESIGN

At the same time as these Greek and Roman developments, solar architecture and city planning evolved in China. Like the Greeks, the Chinese had no adequate heating system. They varied the number and thickness of their garments according to the season.[7] This saved large quantities of fuel.

The winters are cold and sunny in much of northern China. Orienting homes to the south made practical sense. They were warmer in winter and cheaper to maintain.

Figure 3. 1897, Peking, China. View of the south facade of a typical Chinese residence. Notice the extensive lattice windows.

The Chinese developed an awareness of solar orientation early in their civilization. Excavations of Neolithic (3500 B.C.) pit-dwellings from Pan-P'o-Ts'un to An-Yang indicate that the crude windows and entrances of these prehistoric houses all faced south.[8] Many historians agree that these ancient peoples oriented their homes south to take advantage of the sun's winter heat. By the time of Confucius—the fifth century B.C.—solar orientation had become integrated into Chinese cosmology. South represented summer and warmth. South was also the direction faced daily at noon when one worshipped heaven and earth. Thus, buildings fronted southward whenever possible.

All the main rooms of the traditional Chinese home opened onto a south-facing courtyard, bearing a striking resemblance to the Greek home. Decorative lattice windows lined the home's entire south facade (Figure 3). By the first century A.D. translucent rice-paper or silk lined these windows, admitting winter sunlight to the interior, but repelling cold winds. In summer extended eaves over the south side of the residence shaded the interior of the home from the sun.

Entire urban areas also fronted south. Ch'and-an, the Chinese capitol in the Han Dynasty (206 B.C.–220 A.D.), represents such a solar-oriented city. All the main streets of Ch'ang-an laid perpendicular to each other, running north-south, east-west. All homes faced south. No house reached a greater height than any other, assuring everyone access to the sun.

In contrast to the west, solar architecture and city planning continued unabated in China from the time of the ancient dynasties up into modern history. As late as the seventeenth century a European visitor reported that "all the cities and palaces of the king, the great lords, the mandarins, and wealthy persons are so built that...the principal rooms look toward the south."[9]

In the seventh century A.D., with the colonization of Japan, China exported her solar legacy just as the Greeks had exported theirs to the Italian peninsula in classical times. The city of Ch'ang-an, in fact, served as the model for Japan's first capital city—Naniwa (part of modern Osaka).

MEDIEVAL EUROPEAN SOLAR DESIGN

In the west, for more than one thousand years after the fall of Rome, European architects and builders virtually ignored the principles of solar fenestration. The writings on solar architecture of the Greeks and Romans fell into disuse. Awareness of the Chinese's use of the sun was relatively unknown. Large panes of glass went out of fashion; they were no longer as available, affordable, or practical.

Window glass became fashionable once again during the Renaissance. This interest in classical civilization led to a preoccupation with Greek and Roman architectural styles. When this classical influence reached northern Europe, however, most architects copied the outward forms of the ancient buildings, but neglected the solar principles that made their special beauty functional.

Often designers faced the classical portico and large bay windows toward the north or other unsuitable directions.[10] Thus numerous wealthy northern Europeans often had to extensively heat their mansions in winter because of this striking misapplication of ancient solar design principles.

NINETEENTH AND TWENTIETH CENTURY EUROPEAN SOLAR ARCHITECTURE

The lot of the urban poor during this era, however, was far worse. The onslaught of the industrial revolution had thrown peasant workers into sordid, sunless living quarters built back-to-back and row-after-row with small randomly placed windows. Such squalid conditions soon bred a host of lethal diseases—cholera, tuberculosis, smallpox and typhoid fever—due in part to the lack of direct sunlight in these areas. Physicians of the period had begun to understand that ultraviolet light destroys bacteria. As epidemics threatened entire cities and even the wealthier districts, reformers, planners, and health officials pressed governments to abandon their laissez-faire attitude toward housing, and demanded that something be done.

By about 1900 many European countries had enacted public health and town planning laws which ensured citizens' sunrights. Legislation passed by the British Parliament, for example, stipulated that rooms and windows must be placed so as to allow sunlight to "enter freely."[11] Germany enacted similar codes. Many working class communities like Port Sunlight and Bournville, England, were established during this time to provide healthier residences for industrial laborers. These communities offered housing with plenty of sunlight, space and adequate sanitation (Figure 4).

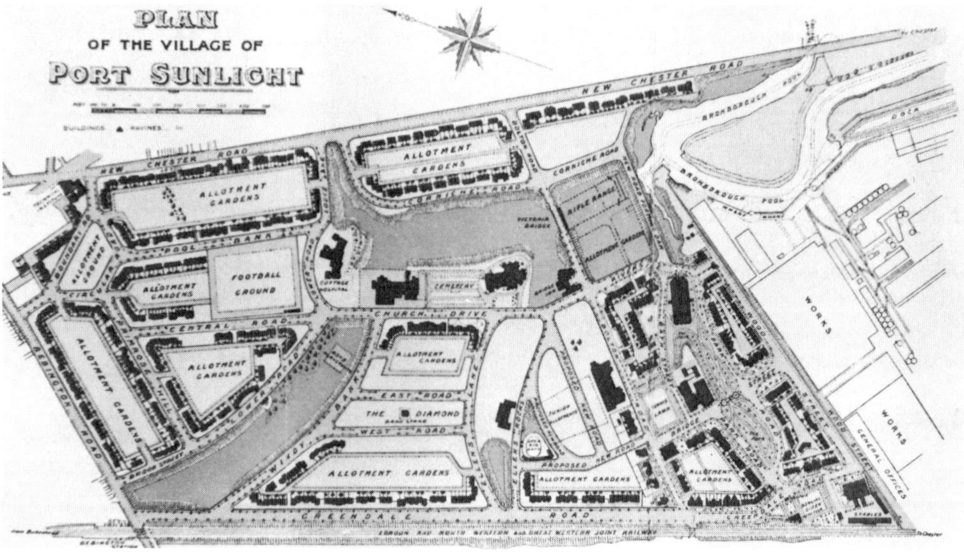

Figure 4. 1916, Port Sunlight, England. This plan of Port Sunlight shows how its homes—unlike those of the common industrial town of the day—were widely spaced apart to allow all dwelling access to the sun. Homes, however, were not oriented in any one direction.

As planned working class communities became more popular in England and other European countries, architects and planners studied the question of solar orientation more scientifically, determining, as the ancients had, that south was the best orientation for a home in the northern latitudes.

Communities designed by these men, however, did not bring about an immediate revival of the highly structured solar communities that were once common in ancient Greece and China. Reacting to the horrors of congested urban slums, which usually had been laid out orthogonally, most planners avoided such grid patterns. This made it difficult to build properly solar-exposed homes.

Most of the spacious, sunlit working class communities that developed in the latter part of the nineteenth century and early twentieth century were built in suburbs or the countryside. The possibilities of extending the same benefits to urban residents inspired Augustin Rey, a French housing official, to investigate the feasibility of urban solar planning.[12]

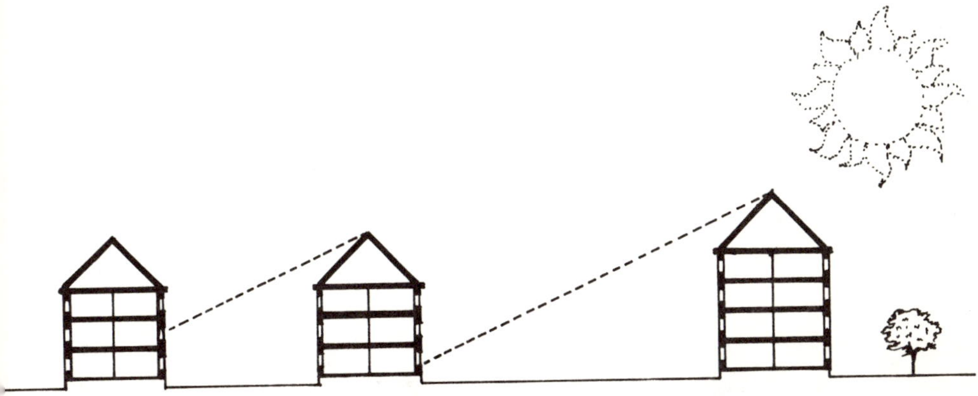

Figure 5. 1913, Paris, France. Sunlight to the apartment on the far left, which is spaced 1½ times its height from the center apartment, is partially blocked during the winter. The center apartment, however, receives all available sunlight because it is spaced 2½ times its height from the building to its south (right).

Rey found that at the latitude of Paris, south-facing apartments built one in back of another must have a space between them of at least 2 1/2 times their height to avoid shading of windows in buildings north of them. But east- and west-facing buildings needed only to be spaced 1 1/2 times their height apart. Therefore, south-facing buildings required nearly twice the amount of land as east-west oriented structures (Figure 5). Extrapolating from his findings, Rey concluded that Paris had nearly three times too many homes for an optimally solar-planned city. He encouraged the French government to adopt rational planning laws which would allow solar planning in metropolitan areas.

Tony Garnier, another concerned Frenchman, attacked the same problem. Garnier, however, envisioned building an ideal industrial city of 35,000 — "City Industrielle" (Figure 6). Garnier designed his city to be composed of long rectangular blocks with all the main rooms and windows of the buildings fronting south.[13] Homes occupied only half of their lot to prevent the shading of adjacent buildings during winter. The remaining land was reserved for private gardens. Although his idealistic plan never really materialized, it helped to stimulate later designers to consider such models.

MODERN SOLAR ARCHITECTURE IN EUROPE

Following World War I, a new phase in modern architecture emerged in Germany. A strong solar housing movement began to flower, and window glass began to be used as a solar heat trap to help keep buildings warm in winter. Germany had to face the harsh economic realities after the war, and according to Marcel Breuer, a pioneer in solar designed homes, saving fuel by designing to use solar heat was a major goal of the German modern housing movement.[14]

Because of the acute housing shortage at this time, large apartment complexes were favored over individual homes. Traditionally, throughout most of Europe, large apartments lined the perimeter of a block—facing in four directions. However, the new architecture consisted of long narrow buildings, several stories high, built in parallel rows. Walter Gropius, a leading exponent of this new architecture, explained that parallel rows of apartments have a great advantage over the old blocks in that all apartments can have an equally favorable orientation to the sun.[15]

Figure 6. 1918, Paris, France. This rendering by Garnier depicts his Cité Industrielle. All homes face south.

Most of the first solar buildings that went up in post-war Germany, however, faced east and west running on a north-south axis (Figure 7). As Rey had calculated earlier, this arrangement was the most advantageous to economical land use. With this strategy long, narrow buildings were erected so that no apartment blocked another's sunlight during the course of a year. The majority of these structures were four stories high and two rooms deep.

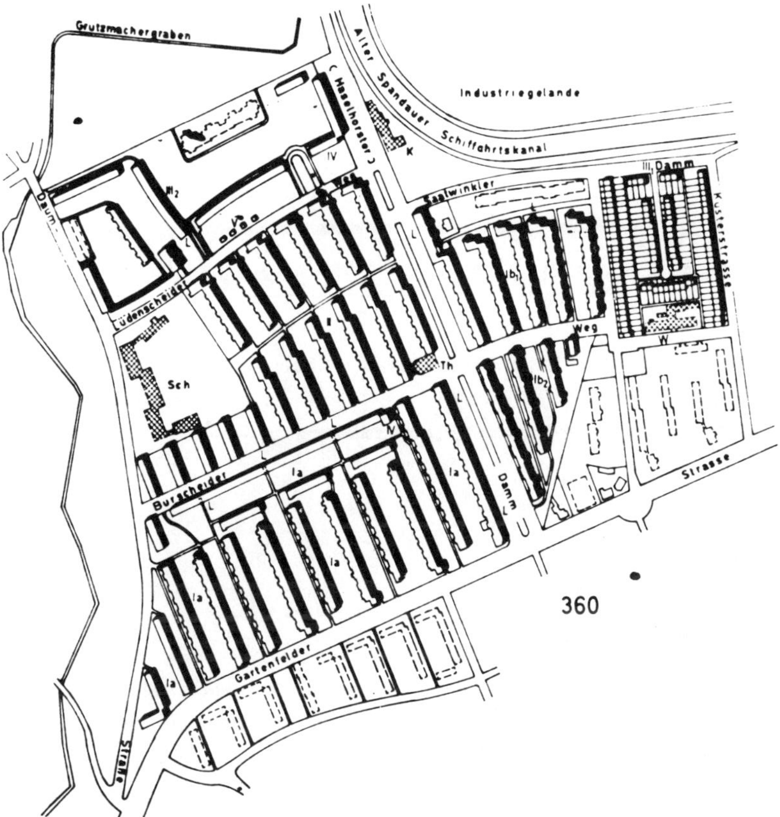

Figure 7. 1932, Haselhorst, Germany. The plan of Haselhorst is typical of early sun-planned German housing developments. North is at the top.

But as the ancients well understood, the winter sun is in the south all day. Thus the east and west windows of these German buildings received only modest amounts of sunlight on winter days in northern latitudes because the sun's rays strike at oblique angles. In summer, just the opposite occurred—the intense rays of the morning and afternoon sun came straight into the east and west-facing windows.

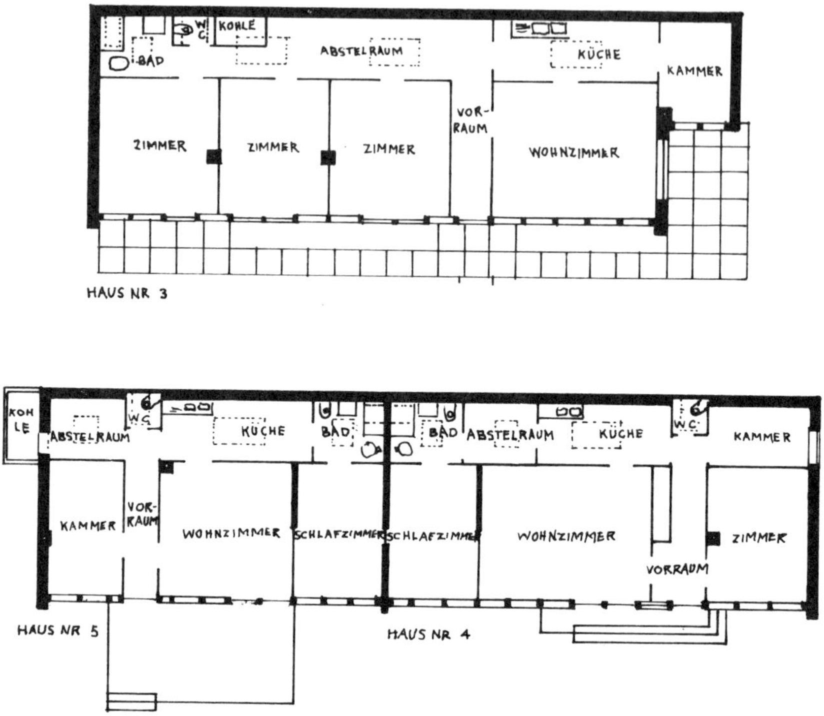

Figure 8. 1932, Berlin, Germany. Häring's floor plans indicate all the main rooms of the house on the southside (bottom) and the secondary rooms to the north (top). The south facade was over 80% glass.

The flaws of this type of orientation soon became apparent, and architects began designing apartments with southerly orientations. But another problem presented itself— for south oriented apartments to be able to take full advantage of the winter sun they had to be very shallow in depth. This reduced their overall economics. In addition, many German architects had already begun to believe that large-scale apartments were inherently uneconomical forms of housint for reasons of upkeep and initial cost.

By the early 1930's, designers began considering single-family homes which unlike apartments, could be standardized and mass produced economically. Many German architects therefore began designing communities composed of rows of solar oriented single-family dwellings.

Berlin architect, Hugo Haring, helped initiate this movement to single-family homes in the 1930s. Haring's homes all took advantage of south orientation, for as he put it, "There is no doubt any more that we, in the interest of obtaining proper solar radiation, must choose a south orientation"[16] (Figure 8). Large plate glass windows along the south facade of his homes admitted generous amounts of winter sunlight, while very small windows on its north side protected the home from cold winter winds. In summer retractable awnings kept the high summer sun from the home's interior.

Unfortunately, this new movement had barely begun when economic and political upheaval led to the collapse of the Weimar Republic and the rise of Nazi Germany. Hitler feared organized workers' communities, most of which were solar designed, because such urban centers made left-wing organizing easier. The Nazi party instead supported isolated and randomly planned communities of traditional German housing. This squelched any further expansion of solar planned communities in Germany.

But solar developments continued in Holland, Sweden, Switzerland, and elsewhere in Europe up until the war. After World War II, growing prosperity and cheap fossil fuels for heating dampened the impetus toward any further solar development.

EARLY AMERICAN SOLAR ARCHITECTURE

The story of solar architecture in America began with its indigenous heritage. Very sophisticated solar communities had been built by the Indian tribes of the American Southwest between the tenth and twelfth century A.D.

Acoma, New Mexico, is one of the most sophisticated examples of this solar architecture. Built atop a rocky plateau, Acoma has three long rows of dwelling units running east to west (Figure 9). Each unit was so built that every residence receives full exposure to the winter sun. All doors and windows open to the south and walls are of adobe. In winter, over one-third of the sun's available heat reached the interior of the home through these walls and windows, while in summer only one-quarter of the available solar heat makes its way through the home's well insulated horizontal roof.[17]

Settlers in the eastern United States were not ignorant of solar building design, especially in the rural, frontier settlements of colonial and post-revolutionary America where many dwellings fronted south. In New England, the colonists commonly built "Saltbox" houses facing toward the sun and away from the cold northerly winds. These structures were two stories high and the main windows fronted south. Many saltboxes had vine covered pergolas which afforded the home summer shade.[18]

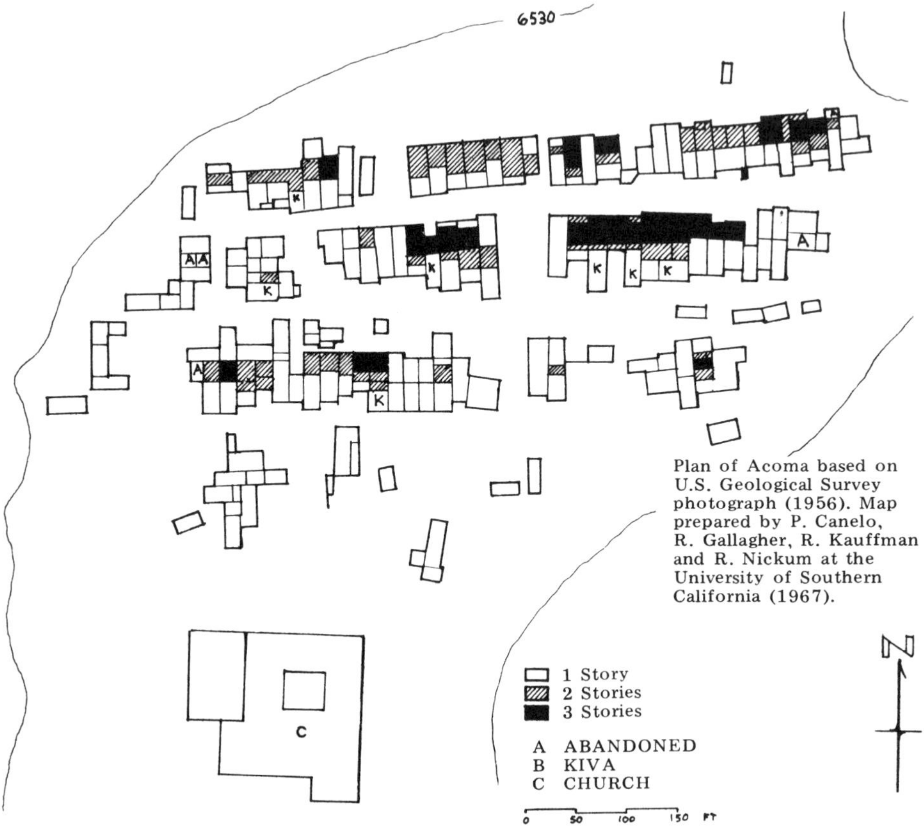

Figure 9. Circa 1100 A.D., Acoma, New Mexico. The rows of apartments in this early pueblo village all faced south (north at top).

As the industrial revolution reached America's shores in the mid to late nineteenth century, formal city plans oriented without regard to the sun prevailed. The common sense solar design adhered to by former rural Americans became difficult to apply in these modern cities and was eventually forgotten.

Around the turn of the century the problem of solar access attracted the attention of William Atkinson, a reform-minded Boston architect. Boston's population had more than tripled in the previous 50 years, bringing a move toward taller and taller buildings. Atkinson saw that skyscrapers enjoyed the advantage of sunlight at the expense of lower older buildings. He developed diagrams that showed how extensively high rises shaded the surrounding buildings and presented his findings to the Boston city council in 1904 (Figure 10). He convinced them that it was essential to guarantee access to the sun, and legislation was soon passed restricting the height of new buildings in Boston.

Atkinson began his work in 1894. He studied the amount of sunlight that penetrates windows facing in various directions. From meteorological data he theorized that east and west windows were the least desirable because they received too much sun in summer and very little in winter. He determined that just the opposite was true for south facing windows.

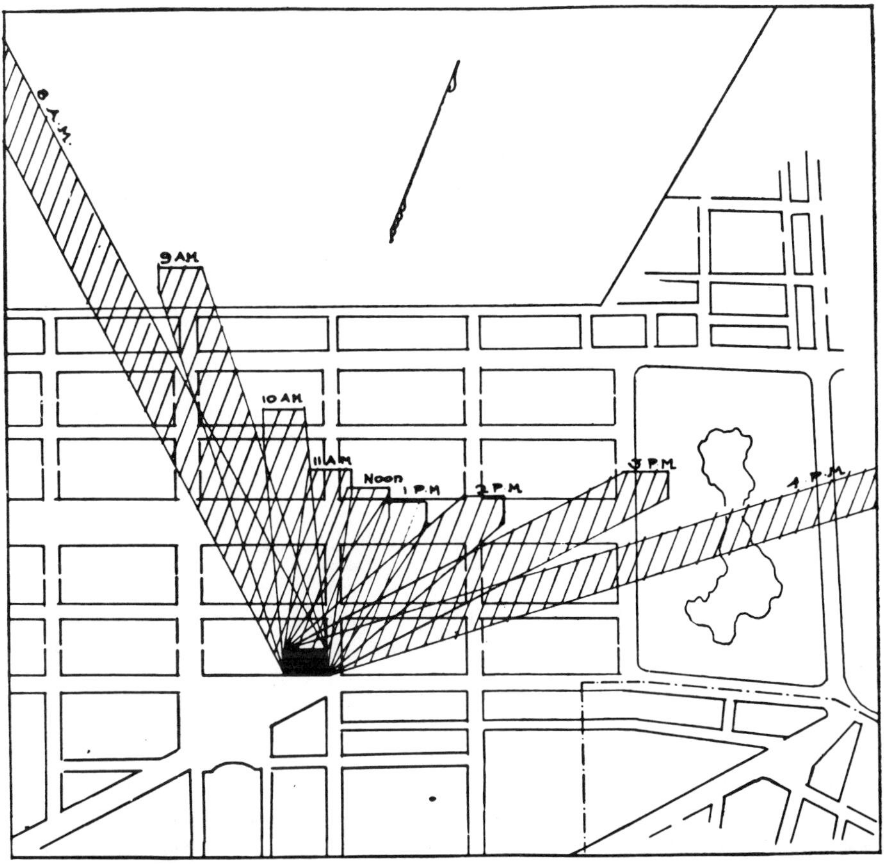

Figure 10. 1904, Boston, Massachusetts. Sun/Shade plan developed by Atkinson. The black rectangle represents a 300 foot skyscraper and the dashed patterns indicate the shadow cast by this building on December 21.

To verify his assumptions he invented a device called the "sun-box"—an insulated den box covered with glass on one side19. He mounted twooden box covered with glass on one side19. He mounted the sunbox vertically so that the glass cover could act as a window, accumulating heat as it would in a room. Atkinson made two such boxes and exposed them to the direct rays of the sun. Between the months of June and December, he changed their positions periodically and collected data for their different orientations—west, southwest, south, southeast, and east.

The tests confirmed his earlier belief: on June 21 an east-facing box reached more than 120F by 8:00 a,m,, 52F hotter than a similar south-facing box, and 42F hotter than the temperature outside. Similar results were obtained in the west facing box. The most spectacular results, however, occurred on December 22 when the temperature of the south-facing box rose to 114F and it was only 24F outside. Amazed, Atkinson wrote, "The sun's rays are not of indifferent value in the heating of our houses."[20]

As a result of his experiments he constructed what he called a "sun-house" in 1912 to demonstrate his newly discovered principle of sun heating. Actually a small shed, the building was put up at Canton, Massachusetts. It was long and shallow, with glass spanning the entire south face so that the sun's rays could fully penetrate the interior. The remaining walls and roof were thoroughly insulated to retain the accumulated solar heat. Despite the freezing weather Atkinson's sunhouse obtained temperatures of 100 F and over with outside temperatures often below zero. This dramatic success of his sunhouse led him to exclaim that "every dwelling could be converted into a sunbox."[23]

Optimistic about the potential of solar heat, Atkinson published a book that same year on his work in solar design. But few American architects took advantage of his ideas and his work was soon forgotten.

MODERN SOLAR ARCHITECTURE IN AMERICA

For two decades after Atkinson published his book, solar architecture lay dormant in the United States. But the progress made in Europe during the years after World War I helped to stimulate a new wave of interest. A number of European reports on proper building orientation began to appear in the English language. The most influential of these was a study by the Royal Institute of British Architects (R.I.B.A.) in 1931-32[21] The institute published a clear, easy-to-follow reference manual on the sun's daily and seasonal movements and the number of hours a day that its rays would strike windows facing in various directions. They also invented a simple device called a *heliodon* that enabled planners and architects to determine the sun's effect on building designs while they are still in the planning stages (Figure 11). By mounting architectural models on a rotating drawing board and using a fixed light source to simulate the sun, designers could easily see what kind of solar exposure a building would have.

Another bit of convincing evidence appeared in 1934. While experimenting with a special type of heat-absorbing glass invented to keep buildings cool in summer and act as an adjunct to early air conditioning systems, the American Society of Heating and Ventilating Engineers (A.S.H.V.E.) accidentally made an important discovery. They were using two laboratories, one with a south window made of heat-absorbing glass and the other with a

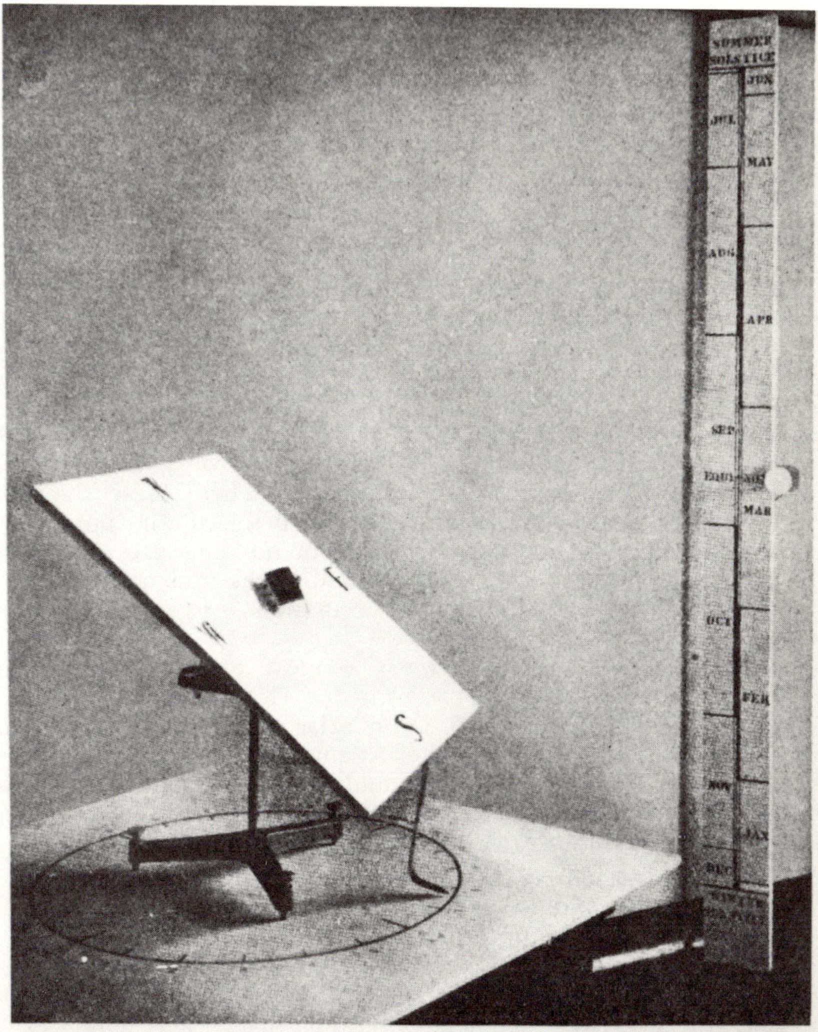

From The Journal of the Royal Institute of British Architects

SUN PLANNING WITH MODELS

The model is placed on a flat board so tilted as to make the required angle of latitude with a vertical axis of rotation. Movement of the board about this axis corresponds

Figure 11. 1932, London, England. View of the *heliodon* developed by the R.I.B.A. A light source on the right illuminated an architectural model mounted on the easel at the left.

south window of ordinary glass. The room with the ordinary window glass received the most winter sunshine, and took nine percent less electricity to heat.[22] The astonished president of A.S.H.V.E. knew that engineers had long accounted for extra heating on the north side of buildings in winter because of cold winds, but these results led him to realize that this extra heat was needed not so much because of the effects of wind but because of the lack of solar heat on the north side.

Two years after this study, the Pierce Foundation commissioned Henry (N.) Wright, managing editor of *Architectural Forum* magazine to determine exactly how much solar heat could be gained by buildings in New York City during different seasons when its windows were located on various sides. At the time, large amounts of glass were being increasingly used in buildings, often with disastrous results. After World War I, factories began to mass produce large, high-quality sheets of glass at reasonable prices. In addition, the development of structural steel eliminated the need for wood or concrete walls to provide support. Now that steel skeletons could be simply covered with a glass skin, architects latched onto the idea of expansive windows and glass walls for purely aesthetic purposes. But indiscriminate use of glass on the north side of buildings led to cold rooms in winter and much greater fuel consumption. If the east or west walls had large glass windows, too much solar gain occurred in the summer.

Wright studied the relationship between window orientation and building heat by applying weather information for New York City to the information published by the R.I.B.A. on solar exposure. As Atkinson and so many others on both sides of the Atlantic had discovered, a southern orientation was best for winter heating and summer comfort. Publishing his results in the influential *Architectural Forum* magazine in 1938, Wright reached thousands of readers with his message.

While these theoretical studies were going on, a Chicago architect named George Fred Keck began putting solar architecture into practice. Keck started his odyssey in 1932 while working on the futuristic "House of Tomorrow," which he had been commissioned to build for the Chicago World's Fair. In an effort to make a lasting mark, the ambitious young architect designed a three-story, twelve-sided building with 90% of its walls made of plate glass. Unaware of previous work in solar design, Keck chose walls of glass for reasons of form rather than function.

One winter day he had an experience similar to Atkinson's which vividly demonstrated how glass traps solar heat. While supervising the building's construction, he found workmen inside shirtless and perspiring. It was 18 F below outside and although the heating plant had not yet been installed it was 85 F inside.

Amazed by the amount of solar heat that had been accumulated in the glass house, Keck, like his predecessor Atkinson, began thinking about using glass to trap solar energy for home heating.[24]

He began developing and trying out his theories of solar heating on homes he designed for private individuals, cautiously putting more and more glass on the south facades. The introduction in 1935 of double-pane glass by the Libbey-Owens-Ford Glass Company helped Keck retain more heat in his solar homes. These windows cut the heat losses normal to single-pane glass by approximately 50 percent. Keck began using double-pane glass in all of his solar homes, in conjunction with over-hangs to prevent the hot summer sun from causing stifling heat inside the home during summer.

The more glass Keck used on the south wall of a home, the more solar heat he found a home would collect. After seven years of experimenting, he decided in 1940 to open the whole south side of a home up with large glass windows. Keck built this home for a Chicago real estate developer, Howard Sloan. All the main rooms of the home fronted south, and overhangs protected the large glass window from the direct rays of the summer sun.

Sloan placed the home on display to judge the public's response. Local newspapers dubbed it a "solar home" thus coining the term. The national media also quickly picked up the story. *Business Week*, for example, called it "the newest threat to domestic fuel." The surge of public interest convinced Sloan to build a subdivision of solar homes, and in 1941 he constructed Solar Park—the first completely sun-oriented residential community in the United States (Figure 12). Meanwhile Keck and a number of other architects continued building structures using solar windows ranging from single-family homes to public schools and hospitals.

Figure 12. 1941, Glenview, Illinois. Two homes in Sloan's Solar-Park subdivision.

Not everybody believed that solar homes could save on a home's fuel bills, and skeptical engineers sought more exact figures on the amount of energy such homes consumed. In October 1941 Libbey-Owens-Ford, developers of the double-pane window, sponsored a year-long study of solar heating.[25] But technical problems in the home—cracks around doors and windows and other underminable heat losses—made it difficult for engineers to accurately gauge the solar heat gain to the home. Yet, they did conclude that solar heat probably compensated for all of the home's heat losses due to the large glass windows. *Architectural Forum* speculated that if the home had been more tightly sealed, the solar heat gain would have exceeded 16 percent of the home's winter heat load.

Owners of solar homes, however, reported even higher savings. The J. L. Bennetts of Barrington, Illinois, for example, said they saved over $100 a year, or 30 percent of the fuel of their neighbors who lived in a comparable house. Others claimed they saved as much as 38 percent on fuel bills.

While homeowners like the Bennetts could gloat, others had to wait until the war's end to share such benefits. The war had halted most new construction. But many articles about solar homes appeared in both the professional journals and popular magazines during this period. *House Beautiful* claimed that "No idea of the last 30 years has so fired the imagination of the American public as the one of letting old sol reduce the winter fuel bill."[26]

Given prominent play on the pages of popular periodicals, the growing interest in solar homes had a tremendous influence on idle developers who were gearing up for the inevitable building boom that would follow the war's end.

The housing market exploded after the war ended. In 1944, only 144,000 new homes had been built; by 1946, the figure hit a million per year. Many manufacturers in the prefabricated home industry, such as Green's Ready-built Homes and Wickes Inc., adopted solar designs. New companies utilizing mass-assembly techniques, such as Solaray Homes of Boston, also entered the field (Figure 13). One of Solaray's developments in South Natick, Massachusetts, consisted of 16 homes selling for $10,000 each. The houses were site-planned individually, with large bay windows on the south side and eaves to block the summer sun. Solar developments built by others ranged in size and in some cases exceeded 100.

The renaissance of solar architecture renewed the controversy among engineers regarding the effectiveness of south glass windows, however. Continuing claims of substantial fuel savings were countered with contrary reports. In an attempt to answer this and other questions, Libbey-Owens-Ford again sponsored a study of a solar house. From 1945 to

1947, Dr. F. W. Hutchinson, Professor of Mechanical Engineering at Purdue University, conducted the study. He found that one reason for the varying degrees of success with solar homes was the lack of reliable information on the optimum ratio of window to floor area. This ratio, he found, should vary with the climate — if the window area is too large for a particular climate, the house will overheat and solar heat is wasted; if too small, little solar heat is available. Hutchinson developed tables and a nomograph to allow architects to make more accurate estimates of the proper amount of south facing glass for given locations.[27]

Hutchinson also demonstrated that a solar home has much greater temperature swings than an ordinary home — it warms up quickly during the day due to the accumulation of solar heat, and cools off rapidly at night due to excessive heat losses through the windows. Some architects had realized that it was advantageous to protect windows at night with heavy drapes or shutters to reduce heat loss. Furthermore, it made sense to develop a storage system to capture the excess solar energy during the day for release at night.

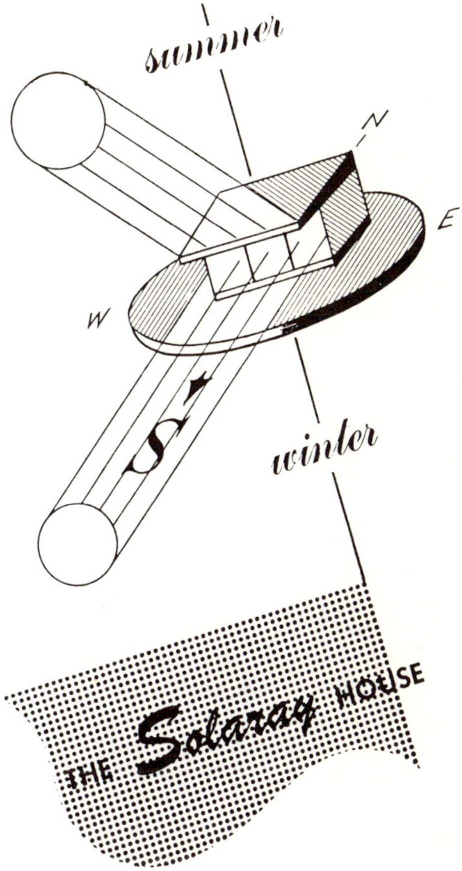

Figure 13. 1946, Boston, Massachusetts. Promotional brochure for Solaray pre-fabricated solar homes.

Even before Hutchinson's study was complete, an Arizona architect named Arthur Brown developed a heat storage system that allowed solar heat to be used at night.[28] In 1945 Brown built a long, narrow building with a south facade entirely of glass (figure 14). A sun porch and dining room occupied the sould inside of the home, and a black concrete wall ran down the center of the home. On the north side of this dividing wall were the home's bedrooms, living room, and kitchen. During clear winter days, sunlight struck the black hallway wall and solar heat penetrated the concrete. Brown had established the fact that heat moves through concrete at a rate of one inch an hour and winter daylight in Tucson lasts about eight hours. So he built the wall eight inches thick, and by evening the solar heat was hust beginning to penetrate the wall and radiate from the bask side into the rooms in the northern part of the house. He also recommended that curtains be drawn behind the glass soon after sunset to retain the solar

Figure 14. 1945, Tucson, Arizona. Interior view of Brown's solar house. The south facade of the home (right) was entirely of glass, and the central interior wall of the home (left) was of concrete and painted black.

heat. Although this heat storage system was never quantitatively tested, it definitely increased the sun's share in heating the home.

Solar building began to falter by the late 1940's, partly because the wartime conservation ethic had swiftly faded. With energy becoming ever cheaper, few cared if the sun could help cut heating bills. People were more concerned with the initial cost of a house, and solar design—due to the expense of double-pane glass and individualized siting—had a price tag up to ten percent more than a comparable conventional house.[29] In the highly competitive housing market, these higher first costs drove customers away or led to lower profit margins if developers reduced prices to attract buyers.

Companies that mass produced solar homes encountered other problems. Most prefabs, including those that were solar, were built with special lightweight wall panels that helped keep construction costs down. These panels did not meet the antiquated building codes of most cities, though, and many companies in the business folded.

The reputation of solar homes also took a beating. The look of large glass windows was attractive to many people, but many did not realize that the purpose of such a design was originally and essentially functional—not merely aesthetic. Consequently, the public readily bought homes designed with improperly oriented windows. Such misapplication of solar design led to disastrous results—over-heating in summer and huge winter heat losses. Many times these structures were advertised as solar homes, leading to a muddling of the term.

Despite these snags sun-oriented homes continued to be popular for a time among the wealthy. They could afford to hire an architect well-versed in the techniques of solar heating and natural cooling. Custom-designed solar homes flourished from Maine to California and from Texas to Canada during the 1940's and early 1950's. Eventually, however, the increased popularity of mechanical heating and ventilation systems—compounded by constantly falling fuel prices—led to an almost universal disinterest in solar architecture by the late 1950's.

PATTERNS IN THE HISTORICAL USE OF SOLAR ARCHITECTURE

Historically, all those who have applied the principles of passive solar architecture have understood the comfort such design can bring to structures. But the use or disuse of solar architecture throughout history has hinged on two major criteria — the prevalent political and social ideology and the availability of fuel.

In ancient Greece, China, the American Southwest and modern Europe, the use of solar design conformed to a prevailing egalitarian ethos which held that all citizens should be able to utilize solar energy. In ancient Rome and in America during the 1930's and 40's, however, class privilege enabled only a relative few to enjoy the comfort and economy of solar design. The concern over fuel in ancient Greece, Rome and post World War II America also motivated many to use solar architectural principles.

Today, we face a question similar to one confronted by many throughout history — will we turn to the sun to augment our dwindling fuel supplies and do so in an egalitarian manner, or will we ignore solar energy or allow it to be affordable only to a relative few?

REFERENCES

The author would like to thank Mr. John Perlin for his assistance.

1. Xenophon, *Memorabilia, VIII*, viii, 8f.

2. Jarde, August, "Note Sur Une Inscription de Delos", *Bulletin de Correspondence Hellenique*, vol. 47, 1923, pp. 301-306.

3. Forbes, R. J., *Studies in Ancient Technology*, Leiden; E. J. Brill, vol. 6, pp. 55-56.

4. Pliny, *Epistulae*, ii, 17.

5. Forbes, R. J., op. cit.

6. *Digest*, 8.2.17.

7. Mirams, D. G., *A Brief History of Chinese Architecture*, Shanghai: Kelly and Walsh Ltd., 1940, p. 3l.

8. Wu, Nelson, *Chinese and Indian Architecture*, New York: George Braziller, 1963, p. 35.

9. Magalhaes, Gabriel de, *The History of China*, (1688), p. 27l.

10. Repton, H. and Repton, J. A., *Fragments on the Theory and Practice of Landscape Gardening*, London: T. Bensley and Son, 18l6.

11. Robinson, C. M., *City Planning*, London: A. Rivers, Ltd., 1909, p. 236.

12. Rey, A. A., "La Ville Salubre De L'Avenir", *International Congress of Local Authorities*, (1913), pp. 217-24.

13. Pawlowski, C., *Tony Garnier: Et Les Debuts de L'Urbanisme Fonctionel en France*, Paris: Center de Recherche D'Urbanisme, 1967, p. 240.

14. Breuer, Marcel (interview with author).

15. Gropius, Walter, *Scope of Total Architecture*, New York, Harpers, 1955.

16. Haring, Hugo, "Bemerkungen Zum Flachbau", *Moderne Bauformen*, vol 33, pp. 619-27.

17. Knowles, Ralph, *Energy and Form*, Cambridge, Mass.: M.I.T. Press, 1974, pp. 28-33.

18. "Two Craftsman Houses", *The Craftsman*, January 1909, pp. 472-80.

19. Atkinson, Wm., *The Orientation of Buildings or Planning for Sunlight*, New York: John Wiley & Sons, 1912, pp. 62-66.

20. Ibid, p. 78.

21. "The Orientation of Buildings", *Journal of the Royal Institute of British Architects*, September 10, 1932, pp. 777-797.

22. Shaver, W. W., "Heat Absorbing Glass Windows", *Transactions, American Society of Heating and Ventilating Engineers*, No. 1018, Vol. 41, p. 297.

23. "News of the Month", *Architectural Forum*, Vol. 79, p. 248.

24. Keck, George Fred (interview with author).

25. "Solar Heating", *Architectural Forum*, August 1943, pp. 3-4, 114.

26. "Can an Old House Be Remodeled For Solar Heating?", *House Beautiful*, June 1945, p. 75.

27. Hutchinson, F. W., "A Rational Basis For Solar Heating Analysis", *Heating, Piping and Air Conditioning*, July 1946, pp. 109-17.

28. Brown, Arthur T. (interview with author).

29. Burns, Frank (interview with author).

The Architect and the Production House

Jeffrey Cook
College of Architecture
Arizona State University
Tempe, Arizona 85281

INTRODUCTION

It is popular to assume that architects and builders have nothing in common. However, they do share a language—the common language of a constructive calling. Unfortunately, they do not always share a mutual respect. A builder's definition of an architect is "someone who thinks about building without building." An architect's definition of a builder is "someone who builds buildings without thinking." The shabby illusion is that architects design public and private monuments—mostly unusable; and that builders provide usable but ugly tracts of building pollution.

Since builders affect the form and performance of much of the built environment many recent government sponsored programs imply that production house builders will passively solve the energy requirements of the country through a new form of energy responsive housing. But why should builders assume that new responsibility? A builder's role is to build and to market his product, not to educate or transform society.

Yet, as society changes its demands, builders must be prepared to meet new standards and respond to changed needs. It is my belief that a new breed of energy conscious and passive solar standards for tract houses is impending. These new age houses, to be successful, must be initiated by builders. But to be successful, they will benefit from the services of passive solar architects. The demands of these new buildings will be too complex without the input of specialized energy designers.

Unfortunately for everyone, the house is simultaneously the easiest and the most difficult subject—it is a building type in which everyone is interested, and in which everyone is both an expert and a dreamer. Each so-called expert has his own standard. The conservative tastes of average buyers are not the same as those of the adventurous and innovative buyers of uncommon houses, and to date most solar houses must be considered rather uncommon.

Tract houses, or production houses, are not the splendid palaces or secretive retreats of most architectural aspirations except in the imagination. Custom houses and semi-custom speculative houses that have no energy aspirations are not in the same league of taste and performance as the median production house. The inclusion of energy efficiency and passive solar standards thus add dimensions to many levels of production house design.

HISTORY

Historically, there have been periods when handsome cities have been the result of speculative housing. The 18th century squares and crescents of London and Bath were the direct result of speculative housing. For example, John Wood, the elder, (1704-1754) was simultaneously an architect, real estate agent, and builder in the transformation of Bath, beginning in 1725. His son continued the tradition and added to the civic success of appealing to popular taste with the construction of the Circus (1754-1764) and the Royal Crescent in 1769. Examples in London include John Nash's Park Crescent of 1812 and, quite different, Nash's Carlton House Terrace of 1827. Today, one trembles at the possibilities if a speculative builder called his tract "the Circus".

CHARACTERISTICS

These old examples of a happy collaboration between builder, borrower and architect display two fundamental characteristics. They are both *monumental* and they are *repetitive*. For today's market there is not much use for monumentality in suburban America. However, regularity and the repetition of standardized elements and components is central to the production economies of tract building. These characteristics must also apply to passive production houses.

Thus, although there are many successful passive solar houses already erected and operating in all parts of the country, they typically are not intended to be replicated side by side. They are not designed to benefit from the economics of serial production. Speculative passive solar building is not production building. And until the production builder can offer passive solar designs *the general public will not have access to solar houses.*

EMERGING PASSIVE CONSTRUCTION

Generally, the rapidly developing construction techniques of the field of passive solar applications have not lent themselves to the serial building and multiple construction techniques of tracts. Even where builders have pioneered speculative passive homes they typically have been in small numbers with a limited amount of cloning. The highly successful solar Village Homes at Davis, California, has been achieved by incremental development and successive improvement of a series of passive strategies. It is interesting that the success at Davis is also due to some role changes. Neither architects nor builders, nor other talents active there, have acted in traditional or exclusive roles. Rather, there has been a new mix of responsibilities and a shift of job descriptions.

Thus, in addition to new designs and new hardware, passive solar production homes will require some imperative reorientation of talents. Excluded from this new game will be architects who run conventional offices and builders who are afraid to change either their models or their method of operation. Both John Nash, the early 19th century London architect, and Davis, California, the late twentieth century energy conscious town, have succeeded by finding new ways of making whole buildings happen—not just by finding new building plans, new products, or new advertising.

During the past year and a half, I have been involved in two separate design and consulting activities related to major passive programs aimed at builders: in the southeast at TVA and in the southwest for the cities of Phoenix, Tucson, and Yuma, Arizona.

TENNESSEE

The Tennessee Valley Authority is an electric power producing utility with activities spread through parts of seven southeast states. The "solar homes for the valley" program (Figures 1 and 2) grew out of the need to provide a new standard of effective house designs for the region.

Figure 1. TVA Solar House 1. This single level plan with a total of 1200 square feet has 3-bedrooms. A carport on the front and deck on the east side of the house are optional and can be varied to suit budget or site.

The intent is to change the pattern of energy use throughout the power district. It is a geographic climate that primarily requires winter heating. However, some summer mechanical cooling is typically necessary. The use of a heat pump as an auxiliary for comfort backup to passive solar construction was found to be an economical and rational use of resources. It could provide both auxiliary heating and summer cooling with dehumidification.

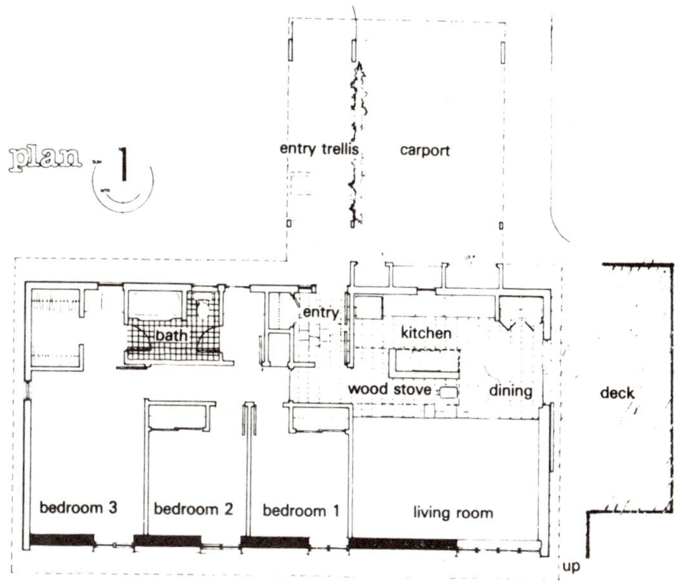

Figure 2. Plan, TVA House 1. The plan organizes principal living spaces along the south wall which provides winter solar heat. Windows contribute "direct gain" while mass walls (unvented Trombe type) provide indirect and time delayed heat. Along the north wall are closets, entry, kitchen, and bath areas that serve as a buffer. This linear plan arrangement allows the house to expand to the east while adding solar exposure to heat the new space.

A total of 11 prototypical designs were developed by a team of designers from the Architectural Design Branch of TVA in Knoxville. Typically, a production house design is a collaborative design effort, as it was with TVA, and there is no single designer to credit or blame for each design. These designs are currently being built throughout the region in a series of diverse climates by a series of builders and sponsoring agencies. A total of 50 will be finished by this writing. They are being instrumented and the winter's performance will be analyzed before the plans are refined and made available to he public.

FIELD TESTING

The simultaneous construction of at least four copies of each of the 11 designs provide field testing opportunities of both planning and construction procedures as well as thermal performance. Among easily solved problems discovered on the construction site are:

- The trombe wall built with mortarless concrete totally violated the modular dimensions intended for the standard sized glazing.

- Methods of venting a Trombe wall during the summer season can be detailed in several different and successful ways.

- The profusion of handles that can result when there are several sliding doors in the same place.

Even before these first stages of field construction and winter testing are complete, there are private builders, as well as materials suppliers, eagerly adapting these new designs into immediate construction projects of their own.

The first two TVA solar house designs have been especially popular (Figure 3). Their simplicity in both appearance and function have proved to be both economical and attractive over a wide geographic area. Their passive solar components are typical.

DESIGN STANDARDS

Architecturally, the TVA "solar homes for the Valley" are aimed for the middle of the market both in terms of budget, and in the style of accommodation (Figures 4 and 5). Because of the large region to be served and the variations of climate and local customs, each design attempts to cater to a particular aspect of this diffuse potential market. There has been a deliberate attempt to establish a conservative style and a non-frightening appearance. The taste of the various designs runs from the 'anonymous' to the careful *Better Homes and Gardens.*

In these prototypical tract designs, the passive elements have been reduced to an elementary simplicity (Figure 6). In order to minimize owner participation, the versatility of manual or automatic controls have been eliminated with some sacrifice in potential heating and cooling performance. Thus, in custom passive solar houses, owner participation is encouraged to maximize the benefit of passive components. But in production houses the owners are only expected to open and close windows and seasonally to open and close the outside vents to Trombe walls.

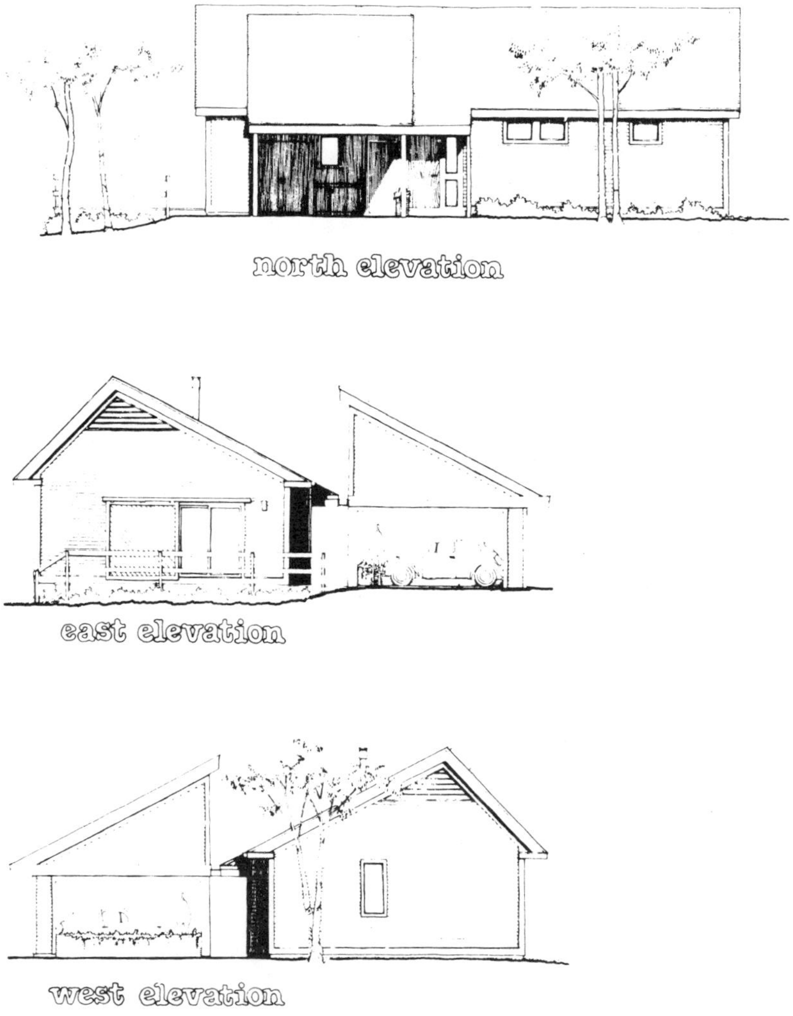

Figure 3. Elevations, TVA House 1. Simple lines are appropriate both to the economies of production building and to marketing. This solar house is designed with the entry on the north side next to the carport.

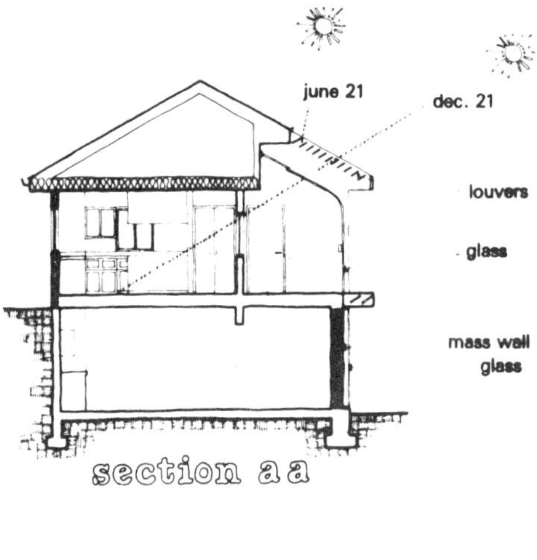

Figure 4. TVA Solar House 2. The entire south wall of this passive tract design is a solar collector. Part of it is window, part is greenhouse and part is a mass wall to provide passive time delayed solar heat. Note the overhangs to keep summer sun off all solar collection surfaces.

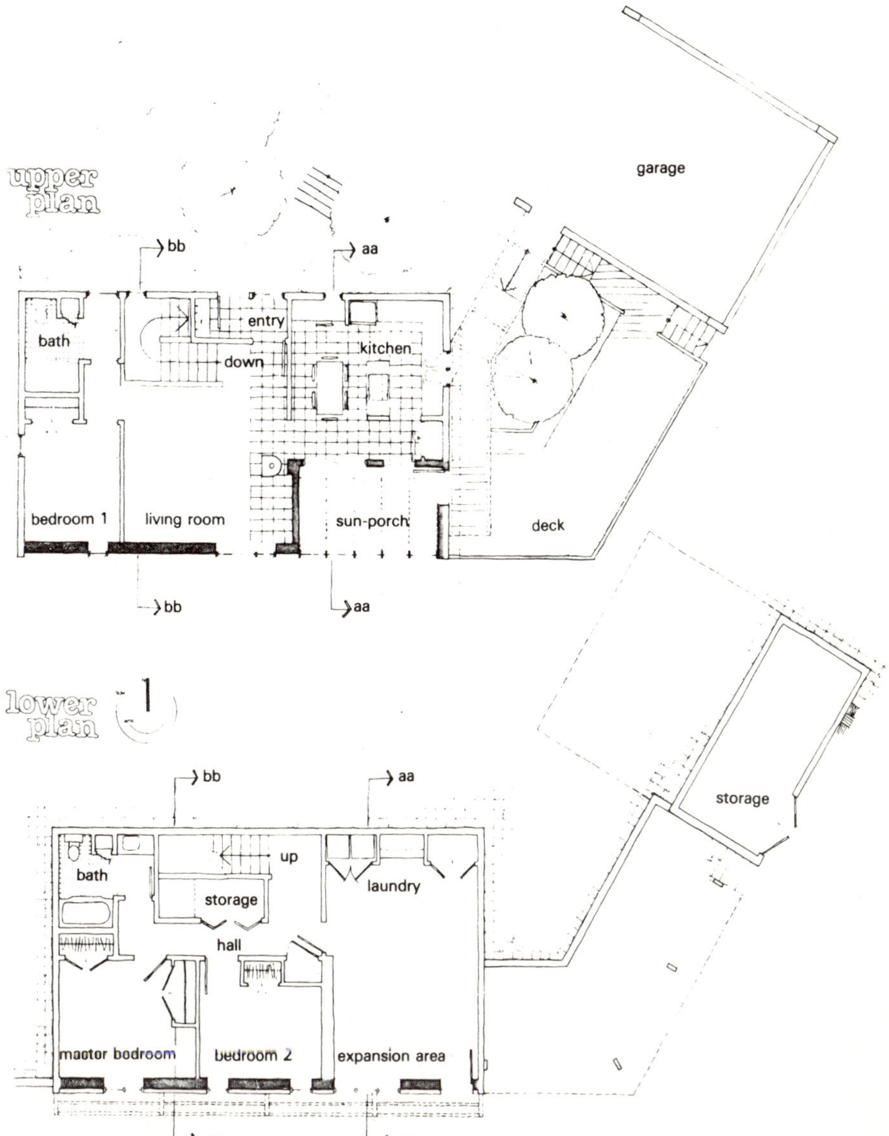

Figure 5. Plan, TVA Solar House 2. A two-level plan with 1008 square feet on each level. 676 square feet are finished on the lower level with 332 square feet available for future expansion. A carport and deck on the east side of the house are optional and allow adjustments in siting and marketing.

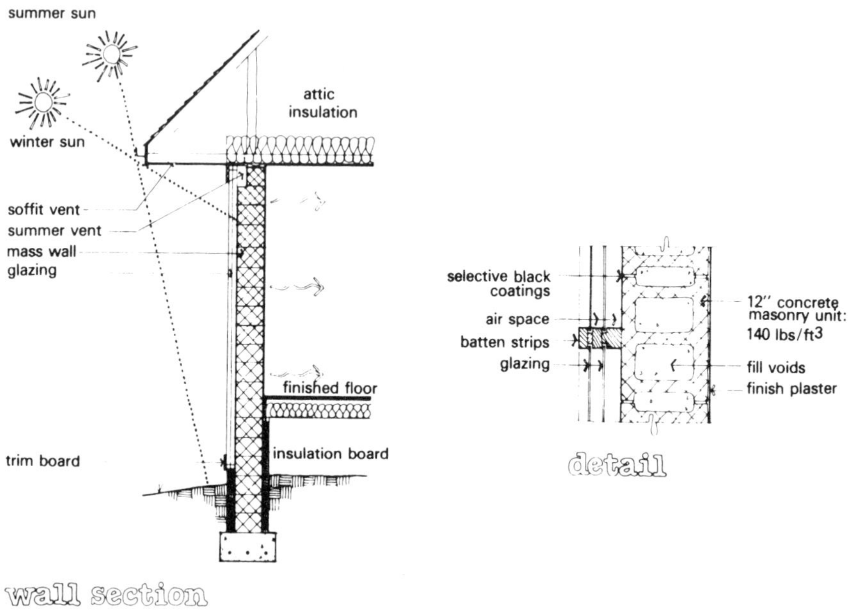

Figure 6. Tract Passive Heating. Solar heat is collected by the "mass wall" which provides approximately 70 percent of the total heat requirement in TVA House 1. The "mass wall" is a simple passive system that requires no mechanical devices and minimum owner participation or maintenance. It is constructed of filled 12-inch concrete block. As the low winter sun strikes its surface, two layers of glass trap and prevent heat from re-radiating back to the outside. The heat is then transferred slowly through the wall and radiates into the main living spaces after the sun has gone down.

Therefore, building components and materials are optimized for one select condition and then fixed without controls to avoid any confusion. While such an approach does not produce the largest annual solar contribution it does minimize the opportunity for system faults such as control failure or human error. This seems to be an important concession when the homeowner or user is unknown until after the house is completed.

COOL HOUSES FOR THE SOUTHWEST DESERTS

On the other end of the Sun Belt is Arizona with its large portion of the southwest deserts. Here the sunshine averages 3600 and 4000 hours per year—it is the only major part of the country where the annual cooling load for houses is significantly greater than the heating load. For Phoenix, Arizona, the cooling load is twice the heating load. The excessive sun of this arid climate is rough on building materials as well as human comfort. The appropriate passive solar strategy for builder's houses in such a climate is to avoid the sun. The best was to keep cool is to minimize heating. *Cool Houses for Desert Suburbs* is the name of the concluding publication of a year long study in optimizing heating and cooling—a study sponsored by the Arizona Solar Energy Commission during 1978-1979.

The goal was to improve the energy performance and environmental quality of builder houses for the desert climates of Phoenix, Tucson, and Yuma, Arizona. The assumption was to use existing off-the-shelf building materials and techniques and the present construction delivery system to optimize heating and cooling comfort. The process of improvement is *design*. Included are the consideration of passive, hybrid and active solar heating and cooling, evaporative cooling and other appropriate design and construction techniques—many unique to the region.

DESIGN STANDARDS

New materials or devices not yet in the market place could add to the performance of these desert tract designs. The encouragement of mechanical cooling devices located outside the walls of the house—such as those proposed using presently available evaporative coolers—would best allow new devices to be refitted to existing houses in the future. Thus, new houses would not be obsolete in a few years time.

Beyond these strategies, the next step toward optimizing heating and cooling, and affecting energy and resource use, would require breakthroughs in construction materials, equipment, housing form, life-style, or land use patterns. No attempt was made to propose new life-styles, novel housing forms, changed comfort standards, innovative land use patterns, startling aesthetic treatments or fresh financing or payback programs. All these potentials for change could strongly affect energy performance and environmental quality not only in desert cities, but in production tracts everywhere.

Fortunately there is a local popular preference in the desert climates of Arizona for masonry construction—a preference which has historical roots. It is a material way of tempering the extremes of daily temperature swings through the use of massive construction (Figure 7). But even low cost houses with insulated frame construction demonstrate some adaptability to climate with their white roofs and vented truss spaces. In addition, evaporative cooling is a local mechanical comfort technique which although highly economical is not presently used well. Evaporative coolers were commercially introduced in the 1930's. But neither their design nor their installation standards have improved significantly since then.

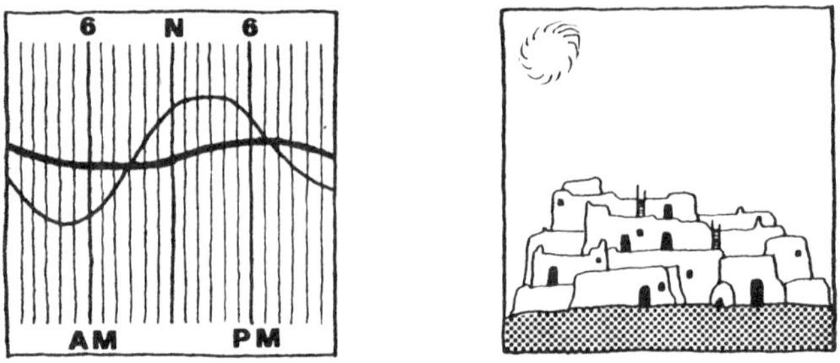

Figure 7. Desert Preferences for Masonry Construction. Traditional building practices in the deserts of the world use massive construction to passively temper the daily temperature extremes.

CLIMATE ANALYSIS

Analyzing typical summer days provided the bases for design. August is the most uncomfortable month of the year in the Southwest desert (Figure 8). Although the days are not as hot as in July, the average relative humidity is higher: the nine hours in Phoenix between 10:30 a.m. and 8:30 p.m. have relative humidities below 35%. The wet bulb (WB) temperature generally remains between 70F and 73F from 8:30 a.m. until 8 p.m., making interior conditions less than ideal for nearly half the day. However, for 13 hours/day complete *comfort cooling* can be delivered by standard evaporative cooling systems on the average day. The use of thermal storage mass built into architectural features could extend this cooling comfort throughout the day.

In addition, there are a whole series of existing strategies that can contribute to the effectiveness of any mechanical cooling. Many of these passive strategies also can contribute to winter passive heating. Thus improving the performance of production house designs is a reasonable and practical goal, well within the means to present budgets and technologies.

CONSTRUCTION STANDARDS

Ideal assemblies for foundations, walls and roof construction were established (Figure 9). A series of standards for windows and solar collector exposure and shading were also defined. In addition, shading standards were proposed which would provide full shade on all passive solar surfaces for at least the six hot months of the year (figure 10). Since the hot season extends into early October such a shading standard is quite different from other locations.

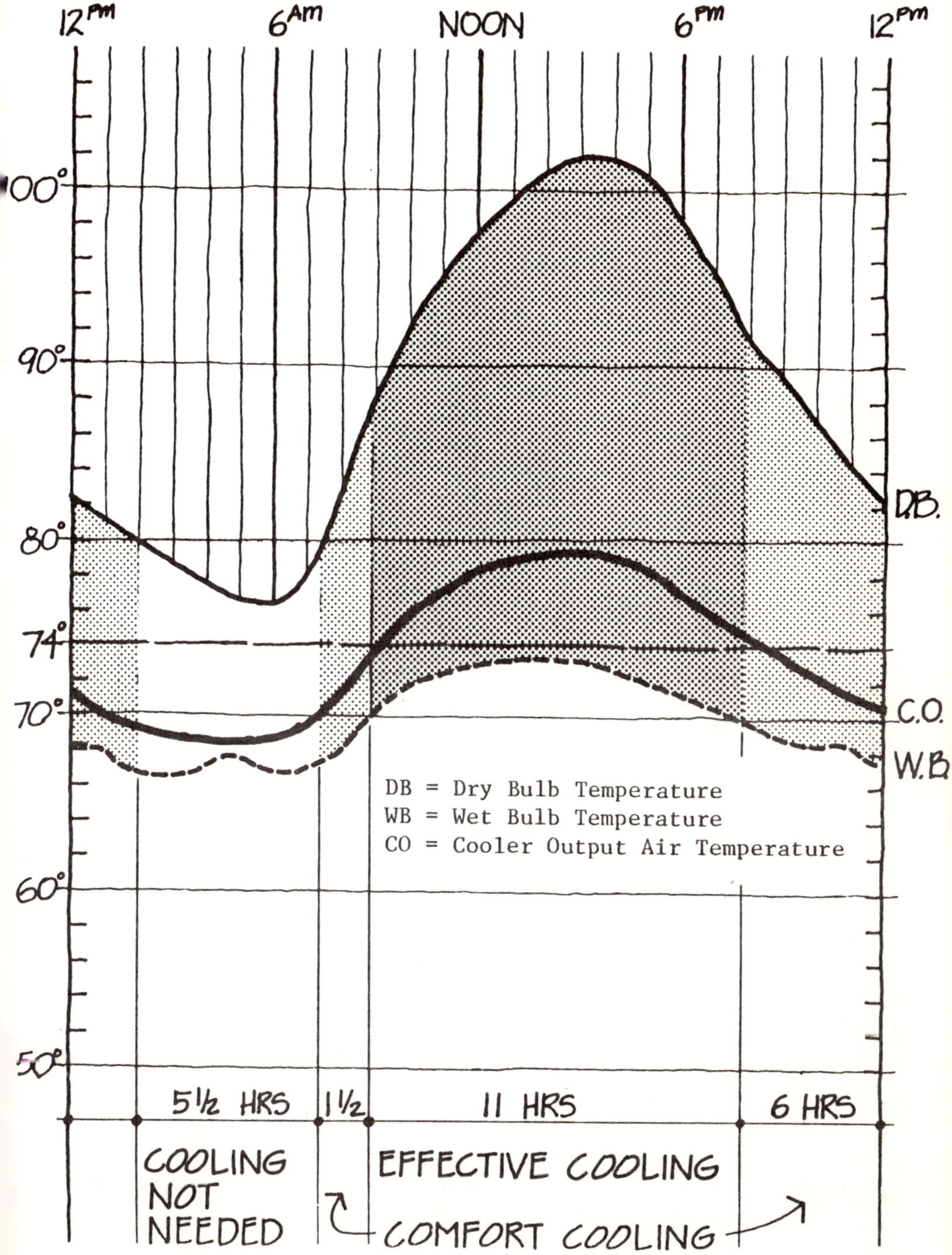

Figure 8. Average August Day in Phoenix, Arizona. During the nine hottest hours of the average August day in Phoenix, the relative humidity is below 35%. Even so the opportunity to achieve total comfort cooling using mechanical evaportaive coolers is still hampered by high temperatures and the presence of enough water vapor to keep the cooler output temperature above 74F. The conclusion for design is the importance of passive thermal storage for cool comfort during the heat of the day.

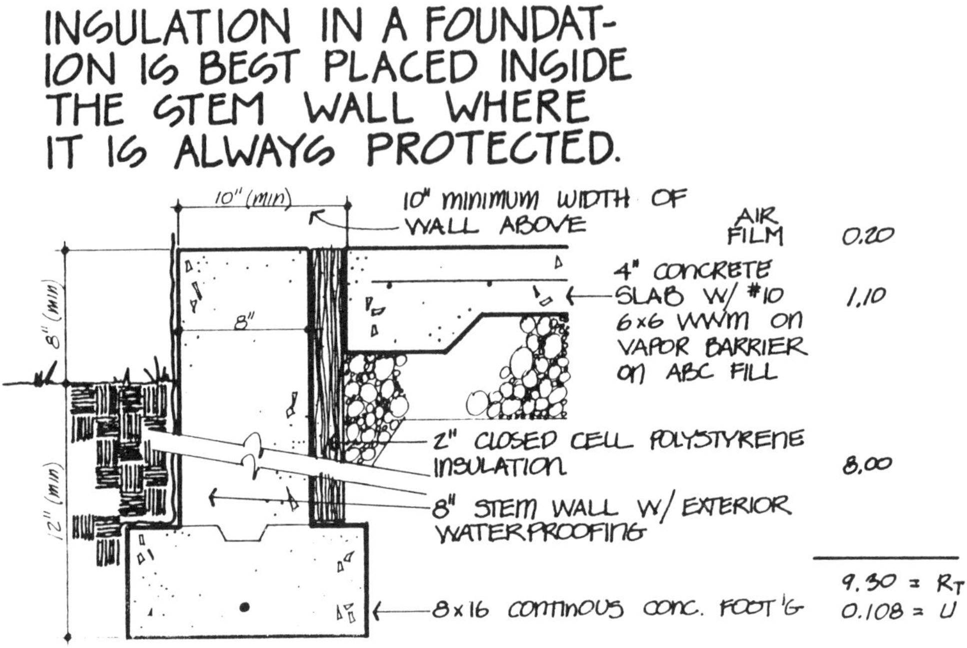

Figure 9. Tract Foundations for the Desert. Local climate and building practice require the development of systems for passive production housing that are different from those appropriate to other climates or construction methods.

After these standards had been established, the largest and most expensive current tract models were examined. Because of their generous budgets, these were the most logical products to test design modifications. On these popular and successful existing houses the program of proposed changes were especially easy to make. Energy calculations verified the improvements. The architectural intent was to make these houses look like they had always looked. There was no reason to change the aesthetics of their appearance. Especially in the popular market it seemed important to make it seem as though no architect had been involved (Figure 11). In general, the architectural goal here was to make passive solar houses look like any other kind of house. Nevertheless in other situations there may be some good reason to give passive tract houses some separate and distinct appearance—marketing strategies must suit the occasion. However, passive solar tract houses do not have to look different.

Thereafter a series of popular plans were redesigned to optimize both their functional spaces and their thermal systems. The most elementary energy strategy for any house is to be sure that the spaces are designed to be

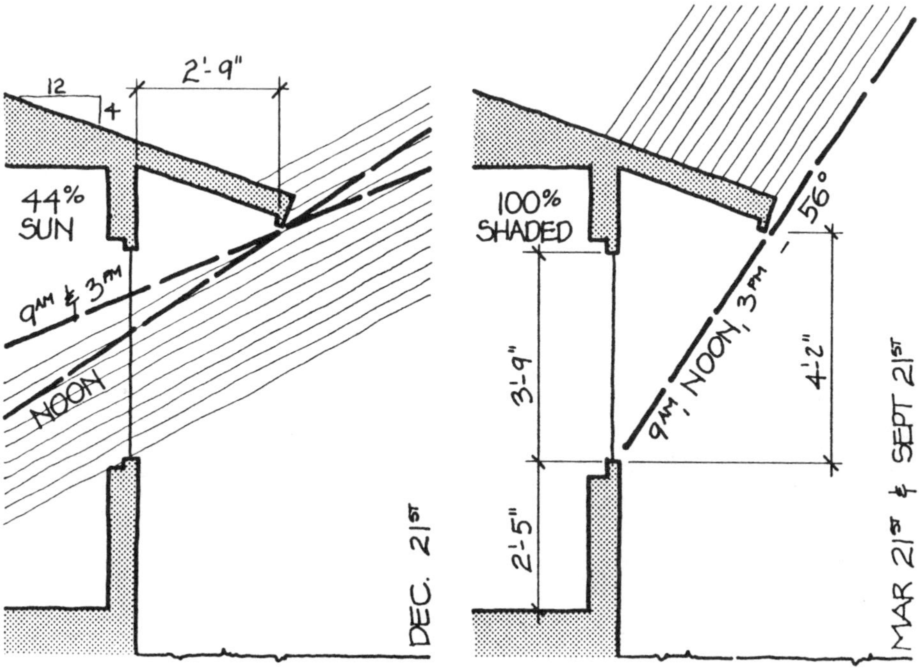

Figure 10. Shading Standards for Desert Tract Builders. When a window, or passive solar surface is completely shaded for at least the six hot months, there is some sacrifice of sun during December. In this case, only about 44% of the window area received sun diring the winter solstice when the sun is at its lowest.

Figure 11. Cool Tract House for the Desert. A popular and successful large, expensive tract design has been modified to enhance its passive performance. This is the south side facing the rear yard. The only obvious clues are the extensive shading of all openings and the generous gable vent louvers.

usable. The smallest house designs with around 950 square feet in three bedroom models were intensively examined. Because of size and budget limitations they offer the most constraints for generous energy redesign. However, optimization of the design still can provide all the winter's heat for the worst January conditions by passive means (Figure 12). House designs were also examined in neighborhood groupings to demonstrate the influence of orientation and zoning on passive energy performance (Figure 13). No serious constraints were found and certain subdivision strategies can reinforce the passive approach.

Figure 12. House 952. With only 952 square feet of living space this tract house has been designed to provide all of its heat for January in a Phoenix/Tucson climate by passive means. Its appearance is intended to be typical and conventional for the local market.

REDESIGN OF THE 1100 SQUARE FOOT MODEL

Perhaps the most successful redesign was the 1100 square foot model. The plan for this modest three bedroom house has several design parameters not immediately obvious (Figure 14). Since it is above the minimum in terms of size and budget, there are a series of options that contribute to this versatility:

- The plan is shown with a thin exterior wall (4 5/8 inches) that provides good insulation, R=19.65, while sacrificing a minimum of floor space.
- The central wall is masonry which provides a continuous bearing line allowing a variety of roof spanning concepts.

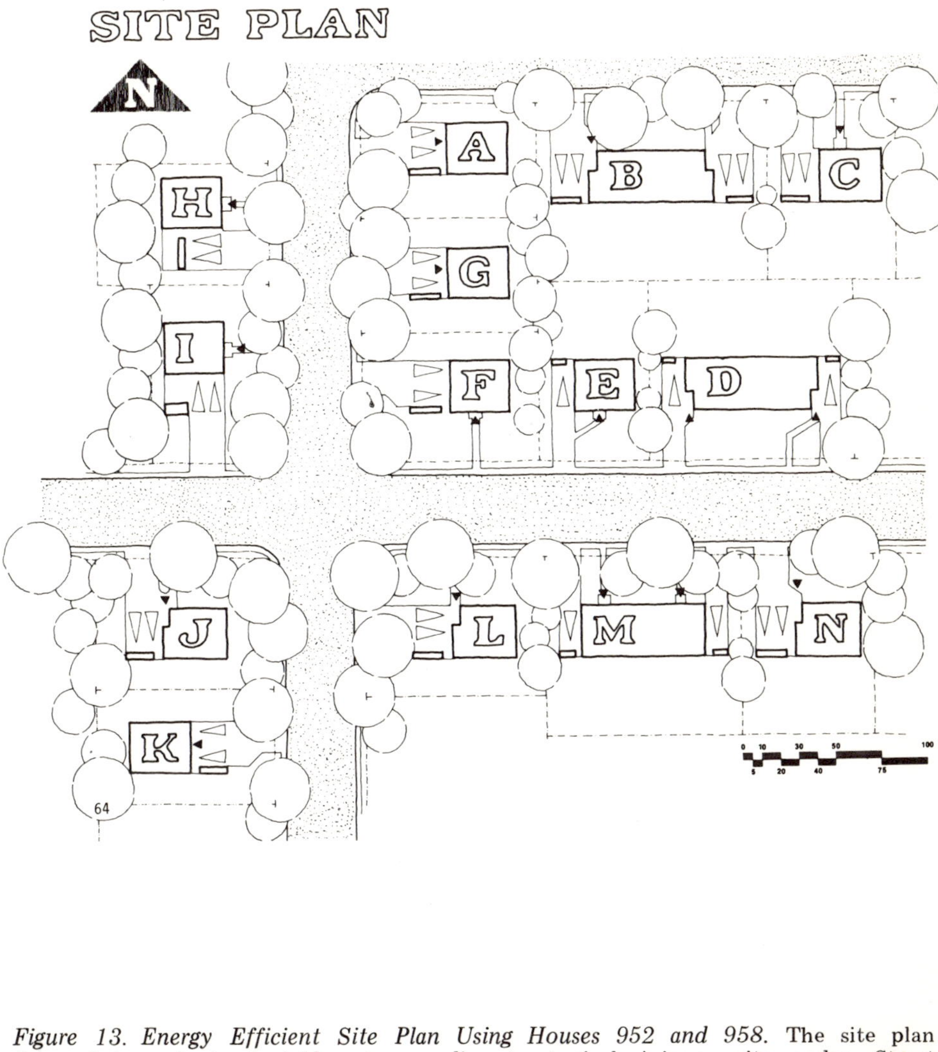

Figure 13. Energy Efficient Site Plan Using Houses 952 and 958. The site plan shows lots and streets laid out according to typical Arizona city codes. Street rights-of-way are 50 feet with 36-foot wide paving. Roll curbs and 3-foot wide walkways are included. House lots are plotted to demonstrate the versatility of plan numbers 952 and 958 while using minimum setback and side-yard requirements. Examples of one and two carports and sample single and duplex arrangements are shown. Each house is oriented to provide maximum solar heat gain through south facing windows during the winter and minimum sun exposure of west and east facing walls during the summer. While east/west streets are preferable for optimum solar orientation, houses "G" and "H" show how the house can be adapted for use on a north/south running street. Corner lots always offer the greatest choice in terms of orientation and entrance. Deciduous trees are suggested on the west and the east sides of the houses and on the south sides of the streets.

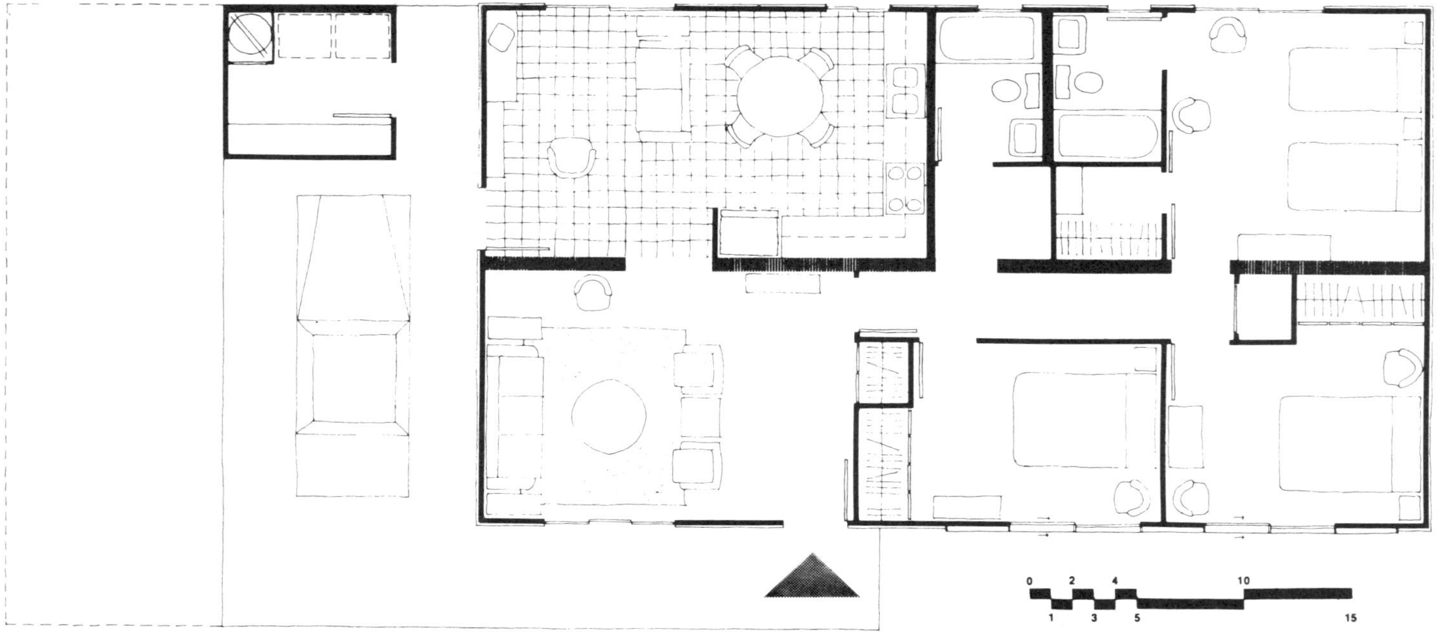

Figure 14. Plan 1100. This house is protected by berming on two sides and the porch roof prevents hardly any direct solar heat gain through the living room window. Heat loss is calculated at 82,000 Btu/average January day. Adding solar collectors between windows above berms adds only 26,000 Btu/day. With net losses of 56,000 Btu/day, this house would require three additional solar collectors (4X8 feet) mounted on the roof (but not shown here) to bring it up to the standard of the other houses. Although this design looks climatically sensitive, it is not. This elevation with the entry to the south is not recommended. By reversing the orientation (placing the entry on the north) the design becomes efficient with a heat gain of 51,000 Btu/January day. Nearly half the south facade is glazed with patio doors while trellises provide summer shade.

- The central masonry wall increases the thermal mass of the house by being exposed on both sides.

- The rectangular outline allows a large variety of exterior treatments but provides an ideal envelope shape (ratio 1 to 1.7) for a desert climate.

- The clear span of 12 feet allows the use of an earth covered roof; or a steel deck roof construction which would permit a heavy roof pond system. If sliding insulation panels were used as part of the roof thermal system the garage/carport width of 22 feet could accommodate the house panels in two layers.

This economical production design, like all proposals, was designed with multiple elevations so that the same plan could be used on both sides of the street (Figure 15). A detailed discussion identifies the performance of each variation. Similarly ducts for evaporative cooling and auxiliary heating were designed with two velocity and two air change standards as well as two different equipment positions.

ORIENTATION

The effect of orientation on solar heat gain was studied for house 1100-C at Phoenix, July 21 at 5:00 p.m. Calculating the summer heat gain of a house facing different directions demonstrates the importance of orientation. Heat gain through the floor slab at the perimeter remains constant, as does the heat added by lights, appliances, cooking, and the occupants themselves.

One would expect a rather long, rectangular plan such as this house has to gain much more heat when the long sides are facing the east and west—open to intense morning and afternoon sun. However, because the long walls are punctured by large windows and the short walls have no windows at all, all sides of the building have about the same exposed solid area. Since the walls are well insulated at R-19, approximately the same amount of heat enters the walls no matter which direction it faces (Table A).

However, the windows are another matter. They contribute most of the heat. Since there are no windows on the short walls, the additional heat gain there is zero. The unshaded windows on the long walls contribute most of the heat gain. Single glazing not only has little insulation value, it serves to trap heat like a greenhouse. Even when white lined draperies are used, windows on the east and west act as very efficient solar collectors in the summer.

As designed, the house could maintain a comfortable inside temperature with the equivalent of one ton of refrigerated air conditioning. But by exposing the windows to sun and by rotating the house with the entry to

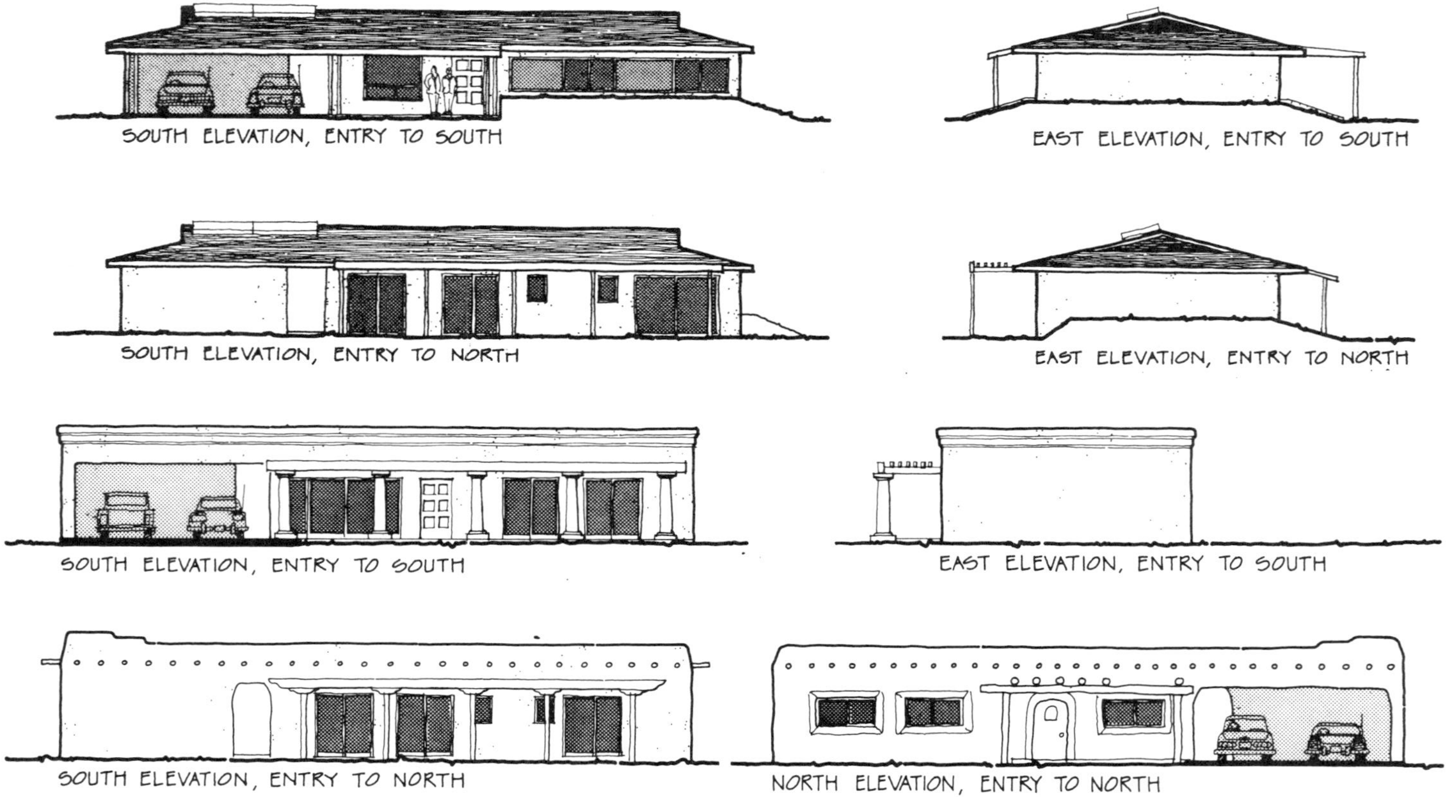

Figure 15. Two Versions of Elevations for Plan 1100. Notice that by using removeable shading devices, the south-entry version of 1100-C can perform very well with a net heat gain of 53,000 Btu/January day. Reversing 1100-C to a north entry has little effect on performance. The heat gain is then 45,000 Btu/January day.

the west, the heat gain increases three times requiring three tons of air conditioning under peak conditions.

TABLE A

Orientation	Roof Area	Wall Area	Perimeter	Appliances & People		Total Heat Gain Btu/hr.	Temp. Rise °F/hr.
Entry to South	1136	1202	176	3000	=	14729	3.7
Entry to North	1136	1114	176	3000	=	15958	4.1
Entry to East	1136	1208	176	3000	=	25346	6.5
Entry to West	1136	1215	176	3000	=	38886	9.9
House as designed with shading:							
Entry to South	1136	1129	176	3000	=	11743	3.1

Figure 16. Plan 2225. Utilizing an east facing entry and south protected windows.

One of the larger three bedroom houses, the 2225/2321, was designed to allow a four bedroom option (Figure 16 and 17.) It was also designed so that its front door could face in any direction while maintaining the plan in an appropriate orientation for the desert climate. East and west facing front doors still allowed windows on the north and south walls. Entry options also on the north and south side means that the buyers could have the identical plan and spaces regardless of the orientation of their lot, thus solving the builder reluctance to offer models with restrictions on orientation.

CONCLUSION

There arė many other important aspects of the problem of tract house design as seen from an architectural point of view which cannot be identified here. Indeed, the termination of the Arizona study with the publication *Cool Houses for Desert Suburbs* is intended to be the beginning of the dialogue of getting some Cool Houses built.

Figure 17. Plan 2321. With a south facing entry and shading protection.

Response has already been initiated. The problem of new energy performance standards from production housing obviously involves new design challenges as well as old ones that can benefit from the contributions of an architect. Successful passive tract houses are more than a technical challenge, design is the key.

Passive Solar Systems and the Mass Residential Markets

Ralph J. Johnson
President
NAHB Research Foundation, Inc.

Today I bring you a message of hope and good news. I say hope and good news because I am sure that all of you share my personal sense of enthusiasm for passive solar assisted homes and light commercial/industrial buildings. I say hope and good news because I'm going to talk about the market realities that will help passive solar. In our enthusiasm, though, we must not take the hard facts of the realities of building and selling homes and light commercial/industrial buildings lightly.

For now, let's look at the factors that will facilitate the expansion of passive solar techniques for homes, apartments, and light commercial/industrial buildings. Then we can realistically plan and act to enhance passive solar applications for the mass markets.

GOOD NEWS POINT 1

If homes were being built today like they were prior to the 1973 oil embargo, I dare say we would not be having this meeting. Then, energy was cheap. Conservation was advocated and practiced by only a few. Active solar systems were the dream of few and even fewer had even heard of passive solar techniques. FEA, ERDA, DOE, the HUD Solar Program, state energy offices, ASHRAE 90-75, energy codes and many other energy facts of life that we take for granted today were not in existence then or were barely emerging.

Then, the economic feasibility of passive solar assisted homes was far less than it is today. Then, it would have been necessary to add both the extra cost of additional energy conserving techniques (ECT) and the extra cost of the passive solar components. Further, except for what I am going, for the moment, to call "far out" designs, the fraction of space heating that could be satisfied by passive solar techniques would have been low. Then, expenditures for passive solar techniques would not have been nearly as attractive economically as they are today.

The average level of thermal protection of new homes has increased substantially in the last six years. In our market research work at the Foundation, we obtain a massive bank of data on the characteristics of new homes being built. Responses to a 24-page questionnaire containing over 1,000 characteristics of homes are obtained from 6,000 builders, who account for some 200,000 dwelling units annually. We obtain information on type of home, size, foundation type, and many other general characteristics as well as much detailed information. For example, the question on ceiling insulation has 32 choices plus "other" so it is easy for the builder to respond. The data are much too extensive to dwell on here, but a few examples will illustrate the point.

For the period 1973-1977, R-11 or more (so called full thick) wall insulation increased from 74% to 99% of the homes with frame walls, R-19 or more ceiling insulation increased from 35% to 87% of the homes and double or triple glazing increased from 39% to 68% of the homes. In some of the colder areas of the country, double or triple glazing increased to about 95% of the total market.

Basement and foundation insulation use has increased substantially and foam insulating boards for exterior sheathing increased from 1% to 14% of the homes during that 1973-1977 period. Also, builders have substantially reduced air infiltration with higher quality windows and doors, weatherstripping, sill sealer, caulking and related techniques.

I estimate that in the past six to seven years the design heat loss for the average home in this country has been reduced on the order of 50%. This makes the use of passive solar techniques far more realistic in the market place. In short, builders as a whole are already taking step number one in the design and construction of marketable passive solar assisted homes, namely, providing a relatively high level of thermal protection which makes the economics and comfort performance of passive much more attractive.

GOOD NEWS POINT 2

This high level of thermal protection really works. Some very careful measurements on an Energy Efficient Residence (EER) and Conventional Comparison Home (CCH) for a period of one year, while occupied by similar families, provided some startling results. In a project sponsored by the Housing and Urban Development Department, the Foundation designed, built, and monitored these two homes about 40 miles north of Washington, D.C. The objectives were to try to determine how much energy can be saved using off-the-shelf energy conserving techniques; to find out whether the savings were worth the extra cost; and to learn whether predicted energy savings agree with actual savings.

The results of one full year's measurements, under normal use and occupancy conditions are now in. Here are the overall results: Both homes were located side by side, were the same size, had the same orientation and looked nearly alike from the outside. Part of the experiment was to pick a typical builder's best selling model and, without changing its outward appearance, see what could be done to save energy.

The comparison home was, by ordinary standards in 1976, rather well protected thermally. Both homes were compact one story with half of the lower level below grade. They each had 1,200 square feet on the first floor and 1,200 square feet of conditioned, but unfinished, basement area which was at grade at the rear. They were both rectangular with the south wall of the basement fully exposed above grade. The conventional home had R-11 insulation in the side walls with insulation board sheathing, R-19 in the ceilings, storm windows, storm doors, and R-11 insulation on those portions of the basement walls above grade and two feet below grade. Also it had a very low measured rate of air infiltration, about 1/3 of an air change per hour. Its design heat loss at 9 F was 30,000 Btu/hr.

There were more than 60 ECT's added to the EER home to make its design heat loss about 14,000 Btu/hr at a measured air infiltration rate of about 0.2 of an air change per hour. We used more insulation, heat pump instead of an electric furnace, and a number of other ECT's.

Overall, there was a 70% reduction in energy used for heating and a 31% reduction in energy use for cooling. We reduced energy use for hot water by 49%. This was mostly because we reduced the operating temperature from the normal setting of about 155-160 F to 120 F. This might have had less significance had we not carefully measured the amount of hot water used in both homes. We found that the average usage in the CCH was slightly higher, both totally and in terms of gallons per capita per day, even though the stabilization temperature at the hot water tap was about 35 F higher than in the EER home.

There was an overall reduction including heating, cooling, water heating and all appliances, equipment and lighting of 49%. This was achieved even though the internal loads from appliances, lighting and water heating were significantly higher in the CCH.

I have estimated that 1/5 to 1/4 of the energy required for space heating in February was contributed by passive gain from the south facing glass, the total of which was quite merchandizable and about the same in both homes. On a sunny day with the temperature outside at 20-30 F, the heat pump ceased to operate about 10-12 a.m. and didn't come on again until 6-8 p.m. Even with continuous blower operation, the temeprature on the south facing rooms rose about 5-6 F above that in the other rooms. A minor modification in the return air system would have improved that facet of warm air distribution.

GOOD NEWS POINT 3

It is now possible to obtain some substantial passive solar benefits with little change in the appearance of the home. As noted above, with homes having high levels of thermal protection and in geographic locations having only an average amount of solar radiation, the provision of modest additional amounts of south facing glass can provide perhaps 1/4 to 1/3 of the heating energy required in January and possibly as much as 1/3 to 1/2 of the seasonal heating requirement. This now becomes a significant marketing tool. First, because those are substantial percentages and, second, because in the very conservative mass housing market, radical changes in appearance of the homes are unacceptable.

GOOD NEWS POINT 4

This benefit from additional south facing glass can be obtained for a relatively small additional cost. Up to a point, some passive solar techniques are more cost effective than some other ECT's. Further, as our experience with passive solar homes expands, I believe we will learn how to use the mass and the materials that are already commonly used in homes much more effectively. I believe also that with the much more widespread use of high thermal resistance sheathing, we will find that the mass — typically some 20,000-50,000 pounds up to 100 tons or more — within that envelope can be more effectively used for thermal storage. Also, some very modest amounts of active/hybrid assists, in terms of distribution and storage of heat, may well be advantageous. I think this is so, even though to the purist, hybrid may sound like heresy. In the mass market place however, it is the bottom line that counts.

GOOD NEWS POINT 5

Higher energy prices are not really good news but they are good news for passive solar. I refer to the fact that my more than 25 years concern for energy conservation have positively convinced me that this nation's energy problem is real, that it will probably get worse before it gets better and that it will not go away in the forseeable future. This means that the price of energy will continue to rise on a worldwide basis probably faster, maybe much faster, than the rate of inflation. Thus solar energy will become more and more attractive. The economics of energy will stimulate the use of more and more passive solar features in homes and buildings.

GOOD NEWS POINT 6

Production builders know that prospective buyers are increasingly concerned about the cost for heating and cooling. This is also true for those who pay the energy bills for apartments and light commercial/industrial buildings. The entire industry is searching for ways to reduce the burden of rising heating and cooling costs. This interest on the part of buyers, prospective buyers and occupants will become so acute that they will willingly operate even on a daily basis, if need be, the kinds of devices that significantly enhance the potential for the use of passive solar techniques. I refer to such simple, but daily, tasks as opening and closing thermal protection devices for windows and glass doors, operating shading devices to protect against excessive heat gain and similar items that can contribute much to enhancing the use of passive solar.

GOOD NEWS POINT 7

The use of some passive solar techniques can be incorporated into many new homes without the need to hurdle most code barriers or, as noted above, to try to market homes that have a radically different appearance. Both of these factors significantly enhance the prospects for the acceptance of more passive solar homes in the mass market.

GOOD NEWS POINT 8

I forsee the advent soon of at least several new products that will significantly enhance the application of passive solar techniques. In our daily work we maintain contact with many manufacturers who sell to the housing and light commercial/industrial building industries. I know that a number of these manufacturers are trying to develop marketable new products for homes and buildings that will enhance passive solar usage. Perhaps the single item that will contribute most will be the development and marketing of movable thermal insulation for windows and glass doors that are lightweight, attractive, available in decorator colors and textures, fade resistant, economic and easily adaptable to the hundreds of sizes that we must cope with. This is a very large order. But it is coming and that is good news for passive.

Passive solar water heating equipment and hardware is just beginning to evolve and that is particularly good news for people in the milder climates where solar radiation levels are generally higher and where water heating may well represent 30% to 40% of the total energy requirements for the home. Manufactuers are working on fabrics and floor coverings that are more fade resistant. This is a point of serious concern to many women in relation to passive and furniture, drapes, rugs and carpeting. Other product developments are underway which bode well for the increased use of passive techniques.

Let me hasten to add however, that all is not sweetness and light. There are substantial barriers deterring production builders from the use of passive solar techniques. Fundamentally, they boil down to cost, performance and marketability. What is the actual extra cost of passive features? What energy savings can be clearly attributed to passive? With these passive elements, is the whole house saleable to the mass markets? To market passive effectively, it will be very important to carefully document, in a fashion believable to the industry and to the public, the additional costs for the passive features. The economics of home buying, as many of you no doubt know, are critical. One thousand dollars extra cost increases monthly payments by $10 or $11 per month. This requires an added income of the buyer of about $500 per year before taxes. This added cost can be a substantial marketing burden.

Also, it will become increasingly necessary for the industry to be able to document how many dollars can be saved as a result of the passive features.

Buyers, mortgage lenders and appraisers want to be able to relate added cost to added savings. For most effective marketing, the added monthly cost needs to be offset by the resultant savings even in the first year.

There are a lot of economic and financing reasons for these two needs for information. The buyer is least able to afford added cost during the first year of ownership. The highest rate of mortgage foreclosures, and failures of buyers to be able to pay, occur in the first two or three years of occupancy. This makes mortgage lenders, who normally are very conservative anyway, leery of adding mortgage payment costs that are not offset by monthly energy cost savings. If they are not offset, lenders refer to this as adding to cost without adding to value. If the even more conservative appraisers, who establish the selling price of homes for mortgage purposes, perceive that added value is not equal to added cost, it means that the buyer will get a lower mortgage and be forced to make a higher down payment. This is also a very serious limitation in the mass markets.

Then there is the basic question of the marketability of the whole house. The production builder faces this question everyday. Basically, the question is: will the additional cost, even if it is only $1,000, limit the marketability of the whole house? The area of the house probably has not been increased, nor have the number of bedrooms, bathrooms or facilities been increased. Will people pay the added cost? Thus, the builder faces the question of risking the sale of say a $70,000 product by adding items that may increase its cost only slightly.

Marketability is not only a question of cost, it is also a question of appearance. There has been little change in the appearance of homes in the last 50 years. This is not because the industry is not progressive nor because there has not been change. In fact, there has been much progress and literally hundreds of changes and improvements in new homes. But, they do not look very much different. This is because the mass market is conservative, very conservative, when it comes to the exterior appearance of their homes. Mass marketing of homes that look substantially different becomes increasingly difficult the more the designs deviate from conventional appearance.

There are a number of subheadings under those major hurdles. One of the most important is the question of financing. Early in the new-found publicity about solar, we received many good news reports about mortgage lenders being quite positive in their approach to financing homes with solar features. As experience has expanded, problems encountered, and the economics become more clear, we hear that the mortgage lenders are retrenching to their more traditional conservative stance. Some disturbing reports have been received recently concerning the reluctance of some mortgage lenders to finance housing with active and passive solar features except on a penalty basis — either lower loan to value ratios or higher interest rates. I do not know how widespread this is but it is a situation that needs careful watching. It behooves all of us to do all we can to obtain documentation of cost, performance and marketability that will allay the fears of mortgage lenders and appraisers.

Even so, these hurdles can be viewed as good news because they present a challenge to all of us to sharpen our pencils, sharpen our wits, get the facts and do a better job of providing passive solar techniques for new homes and light commercial/industrial buildings.

Finally, on a very positive note, I believe that within a few years a significant number of volume builders will be featureing passive solar energy assisted homes. They won't really be "passive solar homes." They will be good designs, that sell in their local market, that have passive solar assist features. To accelerate that trend, all of us need to redouble our efforts to utilize the sun's energy effectively.

Issues and Opportunities in Passive Solar Development

Wayne and Susan Nichols
Santa Fe, New Mexico

INTRODUCTION

Passive solar construction represents an interesting opportunity for builders, designers and developers throughout the country. The demand in America for low energy design and construction techniques will unquestionably increase over the next decade. In our market we know passive solar homes work and are salable. They work so well, in fact, that we are convinced passive solar will play a major part of our government's continuing energy initiatives. The opportunity is clear. Our national goals will switch faster than our capacity to design and build low energy structures. Those experienced in these techniques will have a sizable advantage over those just entering the field. The key to the present opportunity in passive solar is an understanding of how the technology fits with residential and light commercial real estate.

This paper will first describe the work we are now doing in Santa Fe. It will then discuss the relation between passive solar and energy conservation techniques in new home construction. We discuss some of the marketing principles in real estate as they effect the acceptance of passive solar structures. We conclude with a discussion of how marketing may affect key decisions in the design, development, financing, and sales of passive solar homes and subdivisions.

Commercial builders in Santa Fe are using more passive solar techniques in custom construction. Susan and I are small builder/developers and are the first in our market to use passive solar as the center of a development program for an entire project. The project, La Vereda, is 19 passive solar homes on 10 acres within the city of Santa Fe.

We have just completed First Village, a luxury solar development of 8 homes on 40 acres, 4 miles south of Santa Fe. Passive solar has played a dominant role in our design, construction and marketing plans. We feel that our success to date is a result of properly mixing passive solar development with open space site development, water conservation, quality construction, and a general business philosophy of environmental concern and responsibility.

The 8 home First Village was started four years ago as a laboratory to test solar acceptance in a real market situation. All 8 homes are sold from prices of $95,000 to $190,000 with house construction costs in the $50 per square foot range. In 1975 very little was known about passive solar design and construction. The high priced custom homes in the First Village project allowed us to experiment with very little risk.

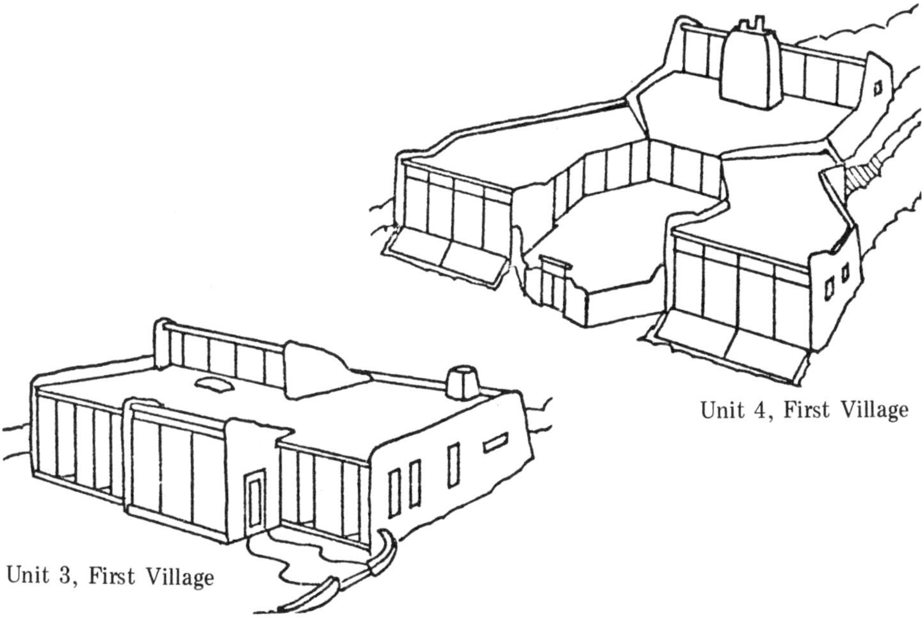

Figure 1.

Our next solar development "La Vereda" is an in-town R-2 planned unit development with 19 homes in three small solar villages or clusters with solar and water conservation required by deed covenant. The unit prices will be $95,000 to $130,000. The sales price, exclusive of land, will be in the $42 to $45 per square foot range. Lot prices are $20,000 to $25,000 for ½ acre of land.

There is substantially more risk in pioneering passive solar in this somewhat lower cost urban market. The trick will be to preserve that important sense of quality, consistency, and responsible development at these lower price ranges.

Passive solar, like any new technology, will go through stages of evolution in the market. Our First Village subdivision is the first generation in this evolution. Here, we have proven that passive solar technology works and can be adapted to local construction and market standards.

Our La Vereda P.U.D. is a second generation solar development. The homes will be standardized models with less customizing. Our emphasis will be on production and cost effective passive solar construction details.

The third generation of passive solar development is the multi-family market. We will accept this challenge once the optimum construction details have been identified for cost and thermal performance.

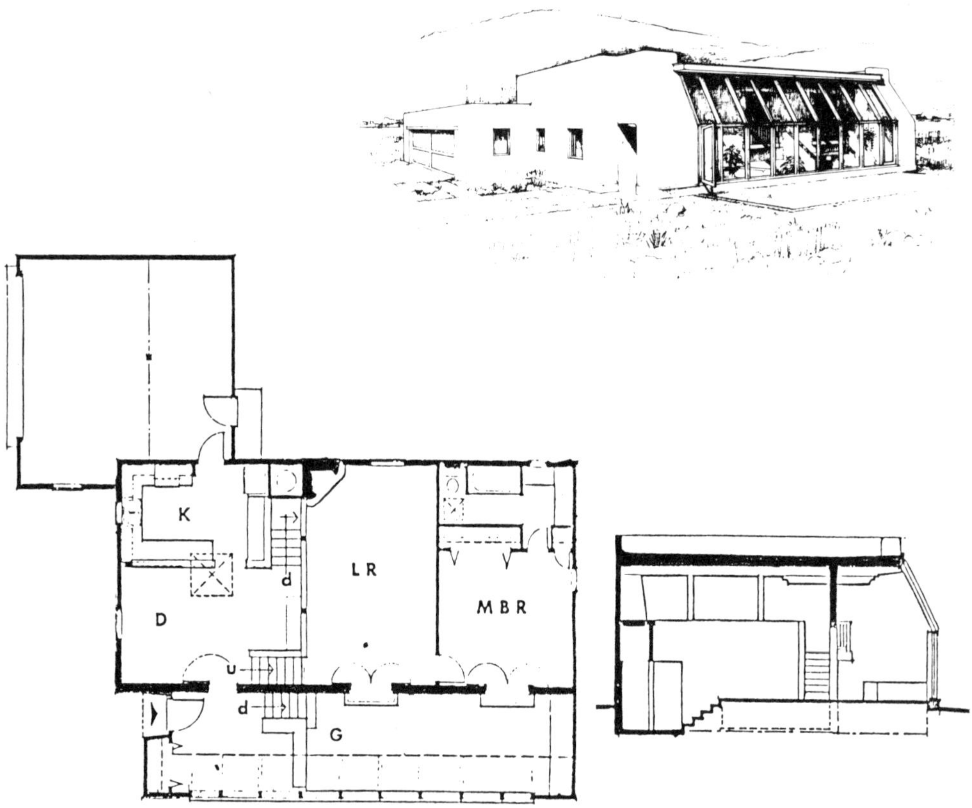

Figure 2. Model 4, La Vereda. Model 4 is a 2,000 square foot split level with 90% solar fraction using mass backed greenhouse and rockbed for $120,000.

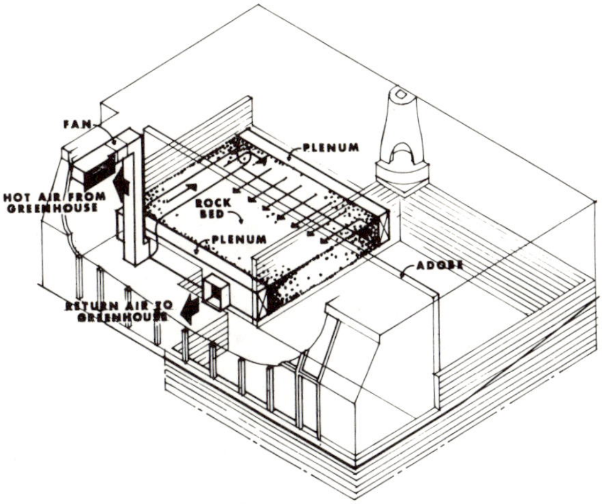

Figure 3. Detail of the greenhouse/rock storage system.

CONSERVATION AND PASSIVE SOLAR

In their book *Energy Future*, Stobaugh and Yergin project 1978 energy use by residential and commercial buildings at 14 million barrels of oil per day or 38% of national energy consumption. They cite evidence that 50% or 7 million barrels could be saved through a combination of conservation and solar energy techniques.

Our experience in building and selling passive solar homes is that once a structure is energy conserving, it makes snese to go ahead and add passive solar. Conservation means extra insulation, air locks, vapor barriers, and double glazed windows. We feel that stopping short of passive solar at energy conservation only misses an important market opportunity.

Doug Balcomb at the Los Alamos Scientific Laboratory has shown that the optimal cost effective strategy for both new and old buildings is a near equal mix of expenditures for both passive solar and energy conservation features. One of the advantages of good thermal envelope design is that it's then possible to utilize passive solar techniques.

At out model-1 in La Vereda, we project heat bills of under $50 per winter with electric back-up heat and a 65F thermostat setting for air temperature. We have no cooling load and the solar heating function is estimated at 87% providing 1.32 million Btu's solar and requiring 1.89 million Btu's from electrical auxiliary heat. This kind of high solar fraction on an energy conserving structure gives these extremely low heat bills. The following table shows that conservation saves a lot of energy, but passive solar drops heating energy cost to a level so low that it creates market interest and excitement.

Figure 4. Model 1, La Vereda. Showing greenhouse and 16 inch cast concrete Trombe wall.

	Passive Solar	Energy Conserving	2 x 4 Built
Total House Costs	$65,629	$64,215	$60,703
Incremental Costs	1,414 (over energy conserving)	3,512 (over 2 x 4)	
Cost per Sq. Ft.	48	47	44
Yearly Heating Bill	28	220	710
Incremental Savings	192/yr.	490/yr.	
Simple Pay Back	$\frac{1,414}{192} = 7\text{yr}$	$\frac{3,512}{490} = 7\text{yr}$	

Table 1. Cost and energy savings for passive solar and energy conservation over conventional construction. (Model 1, La Vereda, study by Bob Taylor, Communico, Inc.)

THE PASSIVE SOLAR MARKET

Passive solar is the extra kicker that gives us an additional sales story beyond just energy conservation. Passive solar matched with energy conservation form a market effective as well as cost effective combination.

Passive solar is a real estate product. It affects value and the use of real property. The long-term potential of passive solar depends on how it fits with overall real estate market trends; and there is no question that reducing energy consumption is a trend in the building business.

The buyer/user choice may be positively influenced by energy and cost saving. If, however, the design, construction, use, and location of the home are influenced negatively by including passive solar features, the overall effect may be negative in the buyer's mind and he will choose another property. The key is whether passive solar complements or detracts from traditional regional conceptions of residential building design and use. The owner or user of a structure is the passive solar consumer. Passive solar will not override the importance of the usual elements of the real estate market mix. It must fit easily into general real estate patterns and preferences within a community.

In the residential market, builders and developers will use passive solar only if it can be proven that it works, it is cost effective, it is acceptable aesthetically to buyers, and it gives the builder a differential advantage over his competitors in the local market. Builders are not quick to adopt new technologies and with good reason. The large amounts of money and lead times involved can make a single mistake financially disastrous.

URBAN MARKET PATTERNS AND ENERGY POLICY

Management of energy use has been a federal responsibility. The national level program has generated substantial policy and research. We now know where we want to go and how. But the actual trip will by necessity have to take place in American communities and neighborhoods. The move from the planning to implementation of reduced energy use requires a shift in focus from the national level to the community and its neighborhoods. Consumption of energy is a basic component of our culture. Waving the magic legislative wands in Washington will not dig into the roots of our attitudes and behavior to produce change. We live in houses, drive cars, shop and work in our own communities. Any change must take place at this level. It makes sense that programs to reduce energy use must engage and motivate us at this level. A plan for community and neighborhood energy management is necessary and possible now.

How should we organize a community level energy program? Neighborhoods, the buildings, transportation, and services that make up our cities have several interesting dynamics. First, the commmunity level is the dominant scale at which we live, work and organize ourselves. For any change to be real and lasting it must manifest and sustain at a relevant scale. Second, our neighborhoods have an interesting and critical characteristic: buildings in a neighborhood were generally built at the same time, and are similar in design size, materials and energy use characteristics. Residents in a neighborhood also tend to share similar socio-economic characteristics. An energy plan directed at neighborhoods would allow targeting solutions, and motivations at the several basic neighborhood types that make up our cities.

Energy consumption is inherent in the structure of our present urban and transportation system. Reducing energy consumption means working with the basic urban fabric and its market pattern. A clear problem with stimulating passive solar and conservation is the fragmented nature of the real estate market. Our building stock involves millions of unorganized, poorly informed decision makers within a highly decentralized milieu where market imperfections dominate, large amounts of capital are required, and buildings have long lives. It would serve our national interest better to develop conservation and passive solar incentives that relate to identifiable real estate markets within cities. There are three broad bands of urban market types, each involving distinct social, economic, use and building design differences. Each of these urban patterns or markets is made up of neighborhoods. Programs tailored to neighborhoods within urban markets would give focus and force to energy use reduction within the myriad of consumers. The three areas are the suburbs, the mid-urban transition area, and the central urban core.

New low-rise construction dominates the middle and upper-middle class suburbs. Here we need federal, state and local incentives for laying out and financing new subdivisions that accommodate solar homes and anticipate future low energy transportation systems.

The mid-urban transition zone is predominately rehabilitation construction of existing low and mid-rise buildings in middle class neighborhoods. Here we need financing and standardized design and construction solutions for solar retrofit and conservation. Transportation and neighborhood organizations are in place and still functioning.

The central city is mid-to-high-rise renewal and rehabilitation construction. Here neighborhood organizations must be created and connected into transportation. There are presently many HUD programs that focus on the central and mid-city. What is needed is an Energy Conservation District zoning that would involve the neighborhood and make a specific area elegible for special rehabilitation, renewal and planning funds for energy conservation and solar utilization within the qualifying area. Present HUD programs for each of these broad urban markets must be coupled with energy conservation and passive solar incentives to provide the necessary planning, organizational, financing, and transportation inputs.

PRODUCT LIFE CYCLE

Passive solar is a new product and will follow a predictable pattern called a "product life cycle". The concept of a product life cycle is helpful in understanding where and how passive solar will be adopted by builders and buyers over the years. The product life cycle starts with a few brave souls who experiment with something new. They are called' early innovators or early adoptors. They do original research and act to model the product for the next group of buyers called middle adopters. The middle adopters rely on the early innovators to test the product and demonstrate its value. The late adopters are the last to use something new and usually wait before

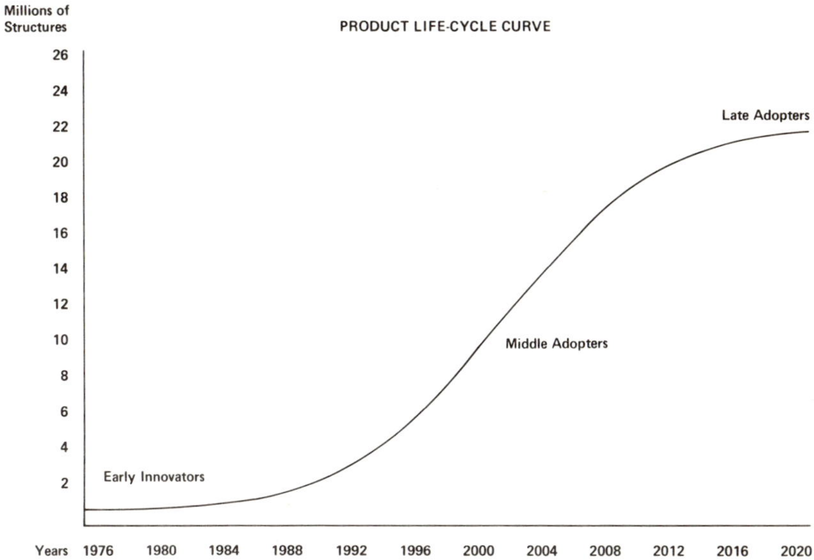

Figure 5. Possible product life cycle curve for passive solar acceptance.

they buy until late in a product's life when prices have dropped. Passive solar has all the earmarks of a successful new product. We are still in the early innovator stage which is why interest is focused on customized buildings involving professional architects. Solar buyers are people who are interested in new things and can afford them. As passive solar matures, we would expect to see the focus shift from custom buildings to tract and mass produced buildings. It is the builder/developer who must carry the passive solar ball through the middle adopter stage of its life cycle.

MATCH THE PRODUCT AND MARKET

Successful selling of passive solar requires matching the real estate product with those buyers who are most likely to buy it. It would be a mistake to develop a passive solar home designed to sell to late adopters. The trick in marketing a passive solar home is to first identify the market segment and then design a product to fit the segment. This is especially difficult in real estate where the product is a mix of location, design, price, construction, financing, and development approach. The successful passive solar home is an interdisciplinary systems problem that must blend each of these interdependent components into a single solution. It is no wonder that builders are hesitant to plunge full bore into the market place with passive solar products.

STATE OF MARKET

There is a natural process of market acceptance for passive solar going on right now. The HUD and DOE programs such as the last Passive Solar Design Competition are doing a great deal to accelerate builder acceptance. In one year there will be hundreds of passive solar residences scattered throughout the country in major markets. These homes are significant because they have been carefully screened by the evaluation teams to pick the best examples of salable, regionally appropriate, high performance passive solar homes. HUD, SERI, and DOE have done a good job in their selection. The infusion of these high quality carefully considered passive solar designs into the market place could accelerate growth of the technology through the life cycle stages.

In plain language, the product could go very quickly from custom homes to the total development stage. We have found in Santa Fe that most of our sales are word of mouth; and feel this has resulted from our taking a leadership role in our market and introducing passive solar first and at a visible scale. We would recommend this to other small builder/developers who would like to establish an innovative, quality reputation in their market. Strategically, the best point of entry into a new product market is just before it moves into the middle adopter stage. At this stage the experimentation has been done and the market is just entering a growth period. If that growth accelerates as the result of an effective federal program, it will be increasingly difficult to achieve visibility in your market. A good plan is to start now with a few homes to test the water, learn from your buyers, and get your local reputation established.

WHO ARE THE BUYERS

Housing magazine called them the hottest segment of the residential market. They are the "Young Professionals". They buy the small car, read magazines, have planned families, are between 35 and 44 years old, and they have the money. In the next ten years, their number will increase from 28 million to 40 million. And, they will control $1 out of every $4 of personal income.

In terms of real estate, the war babies, grown up, represent the primary market for new residential construction for the next ten years. This is a very significant trend. The number of households in the 35 to 44 age group will increase as follows:

	1979	1985	1990
Households	12,236,000	15,538,000	18,224,000
Average Income	**$24,808**	**$27,326**	**$29,485**

Within this group there is a smaller group that is attracted to passive solar. The Franklin Institute has analyzed buyers of solar homes under the HUD program. SERI has done similar studies and they confirm our own sales experience in Santa Fe. We suggest the following profile as the target market segment for passive solar.

1. Age – 30 to 45/ 55 and up
2. Children – 0 to 1.5
3. Income – $25,000 plus (combined)
4. Employment – Two jobs, professional, self-employed
5. Education – Advanced Degrees

Concerns – Quality, Environment, and Flexibility

Figure 6. Passive solar buyer profile.

The literature and our experience indicates another important group which is older households moving to a smaller home and looking for relief from a diminishing value fixed income. Our sales at La Vereda are running about half older households and about half younger couples in the 35 to 44 age bracket.

BUYER MOTIVATION

In our experience, passive solar buyers see themselves as "responsible". This is why they buy a solar home; and they look for the same thing in a builder/developer. They are sensitive to quality and expect the home design, the construction, site, planning, and solar approach to be carefully thought through and consistent with their expectations of a responsible home builder. Part of the appeal of solar is a backlash against what these buyers consider irresponsible builders. The purchase of a passive solar home is, therefore, a personal act with wider social and community implications.

We feel there are four components to the buyer motivations for purchase. They are: (1) Image; (2) Economcs; (3) Comfort; and (4) Information.

Image is a primary motivation for the early solar buyers. To be first in the neighborhood with something new is a strong incentive. The personal and social implications of owning a solar home is also an important motivation.

Economics of energy savings will be increasingly a motive for purchase. The economics of solar heating is still somewhat cloudy. As fuel prices and national concern rise and as subsidy programs are extended to passive solar, the economic benefits will be more apparent and the motivation stronger. It should be noted that passive solar has real economic advantages in most markets. A lower first cost and non-existent operating and maintenance cost for equal performance gives passive solar the competitive edge over active systems in any economic model used.

Comfort is a key to passive solar system's acceptance. A properly designed home with sufficient interior mass, will give good solar heating fractions with very little interior temperature swings over a daily and yearly cycle. The interior mass is basically a radiant heating system without the on/off blasts of hot air from forced air furnaces. In most markets, a radiant system is the preferred high quality heating technique and this is a real plus for the passive solar approach.

Information and control are an important motivation for purchase. Selling the passive solar home is selling information to the buyer about how well the system will perform and the cost savings that will result. We see ourselves as selling a little energy company to the buyer. He wants to know how much extra he is paying and his expected return. The more information we can give, the happier he is and the easier the sale. Detailed solar engineering, incremental cost analysis, and economic projections all help to reduce the buyer's fear of doing something new and untried. Good information is essential to the successful passive solar marketing plan. The basic need that underlies demand for passive solar is the growing sense in America that things are out of the individual's control. Passive solar offers a degree of self-sufficiency to the buyer and increased control over the uncertainties of his immediate environment. Building whole projects of passive solar homes also adds the critical dimension of "community" to this sense of self-sufficiency and control which in turn enhances the buyer's motivation.

CHECK LIST

The following checklist is offered as a guide in putting together a successful passive solar development.

I. REAL ESTATE PRINCIPLES

Remember, good passive solar design will not save a badly conceived real estate project. It will give a good project an important differential advantage in its local market.

a. *Location* is the most important aspect of real estate. Pick a location that will appeal to your target segment. Err on the side of quality.

b. *The Target Segment* should be carefully determined. Who are the prospective buyers in your local market and where in town are they now buying? Prepare a buyer profile to act as the basis for design, pricing and development approach.

c. *A Neighborhood Profile* is also important. Once the location is selected, survey all existing adjoining developments for architectural style, floor plans, price, amenities and types of buyer. It is important to match the target segment with a neighborhood where these types of people are actively buying.

II. SOLAR SYSTEM DESIGN

a. *Architectural Style* is the key to successful solar design. The passive solar elements must be carefully fitted into the styles popular in a market. The better the integration of the passive solar elements with acceptable architectural styles the greater the market acceptance.

b. Passive solar techniques are different for each region. The basic approaches are the same nationally but the mix and proportions vary with local climatic conditions. Survey existing passive solar structures in your region and their performances. Talk to builders and users and select the optimal regional solution. Prepare full engineering calculations and sensitivity analysis for your design and systems and see if the solar fraction is sufficiently high to motivate buyers. Cooling and humidity are problem areas that must be considered. We have found that mass backed greenhouses, Trombe walls, and rock beds give high performance with small interior temperature swings.

III. DEVELOPMENT.

a. *Site Planning* is as important in passive solar development as the house design. We believe that the buyer is looking for a developer who respects and preserves the existing landscape and terrain. The cluster type setting and open space planning using the planned unit development approach allows maximum flexibility for house solar orientation. In La Vereda we have four acres for 19 homes and lots, one acre for roads and five acres of open space.

b. *A Homes Association* is used to own and maintain the open space in all our projects. The Homes Association is also empowered to select and install future energy savings technology if the homeowners so decide. Solar rights are incorporated into the association documents and along with the architectural review are enforced by the association,

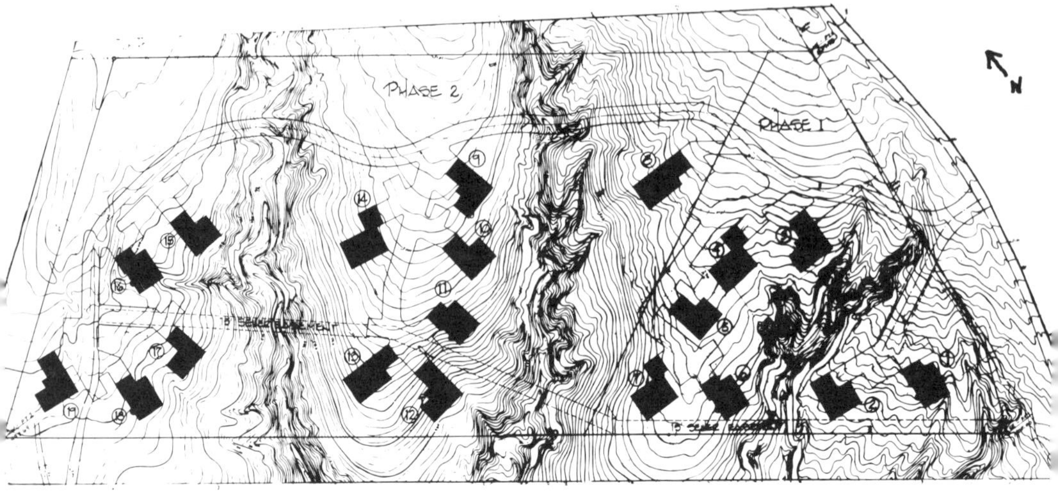

Figure 7. La Vereda site plan.

c. *Community Aspects* to passive solar development help sales. Buyer resistance to accepting something new is reduced if all the homes are solar. Also, the community can give important visibility to a project. By working closely with local zoning and civic groups, a passive solar development is an important community asset. The local utility, university or research laboratory may want to participate in designing and monitoring a project giving it more visibility and credibility.

IV. CONSTRUCTION.

a. *Construction* is the key to good passive solar performance. Infiltration accounts for half the heat loss in a residence. We use our own crews who are experienced in our rigorous quality standards for vapor barriers, insulation installation, and other critical details. Remember, the builder is the solar manufacturer and passive solar is primarily a construction process technology not an equipment technology. Choose a builder experienced in energy saving construction techniques in your area.

V. FINANCING.

a. *Financing* the project depends on project size. Smaller projects of one to six homes can best be financed by local savings and loan associations or banks. On larger projects, it may be advisable to get a permanent loan commitment from your regional Federal National Mortgage Association office. The lender or appraiser is your first sale. We have found it very helpful to spend a great deal of time educating lenders and appraisers on passive solar construction techniques and benefits. It is important to get these people on your side from the beginning. To do this, we prepare an elaborate lender presentation package as an aid in a full-scale lender education program. It is also advisable to separate energy conservation costs like insulation from direct incremental solar costs. Most loan and appraisal forms make this distinction.

Attention to the above issues will help in avoiding the most obvious pitfalls in initiating a passive solar program. Passive solar heating works in most major markets and it can be successfully incorporated into regionally acceptable architectural styles. The size of the opportunity is a function of establishing a local reputation for expertise in energy conserving design and construction techniques. We suggest taking advantage of this opportunity now.

The Practical Side of Producing Windows

Graig Young
Graham Architectural Products, Inc.
York, Pennsylvania

James G. Cribben
Capitol Products, Inc.
Mechanicsburg, Pennsylvania

The practical side of producing windows is an interesting subject in light of the changing market environment where windows are now being considered for their performance as passive solar collectors. To approach the subject of the practical side of producing windows, the market place of today and tomorrow must be examined. The market is changing and it is people who are interested in passive solar energy that will further change today's window market to tomorrow's passive solar collector market.

In discussing the practical side of producing windows, the three key areas to cover are (1) current considerations for window users; (2) performance compromises in the design of windows; and (3) manufacturing considerations.

Today, 99% of the windows sold are being sold as windows—only 1% are currently being sold as passive solar collectors. This will change in the future but the speed which it will change will depend largely on how people throughout the country start to promote and design using windows as passive solar collectors.

With the present market demand being for windows, the architect, designer, or builder should keep the following two practical considerations in mind when designing or building a home.

The first consideration is the use of standard window sizes. All window manufacturers provide window size standards. If the building design is developed adhering to standard window sizes listed in the manufacturer's brochures, the saving in cost to a customer is from 50% to 100% over the cost of non-standard units. In addition, delivery time can be appreciably shortened for standard units because "specials" require processing in a manufacturer's plant double or triple that of standard units.

The second consideration should be to use windows with a tested and certified U-value—not assigned calculated U-values using ASHRAE or other handbook procedures. There are three independent laboratories who test windows to AAMA test procedures and arrive at a measured U-value. When selecting a window for energy saving characteristics, don't hesitate to check for product testing by an independent laboratory and request test reports from the manufacturer.

The second key area in the practical side of producing windows is the performance compromises that manufacturers must make in the construction of windows. One most important compromise, because of its great influence on energy use, is *air infiltration* versus *natural ventilation*. The heat loss due to air infiltration during the heating season has been well documented. The air conditioning load due to lack of natural ventilation during the cooling season is an equally important factor. Unfortunately, in reasonably priced standard manufactured windows, natural ventilation is achieved with the use of a moving sash which loses energy by virtue of air leakage around its seals.

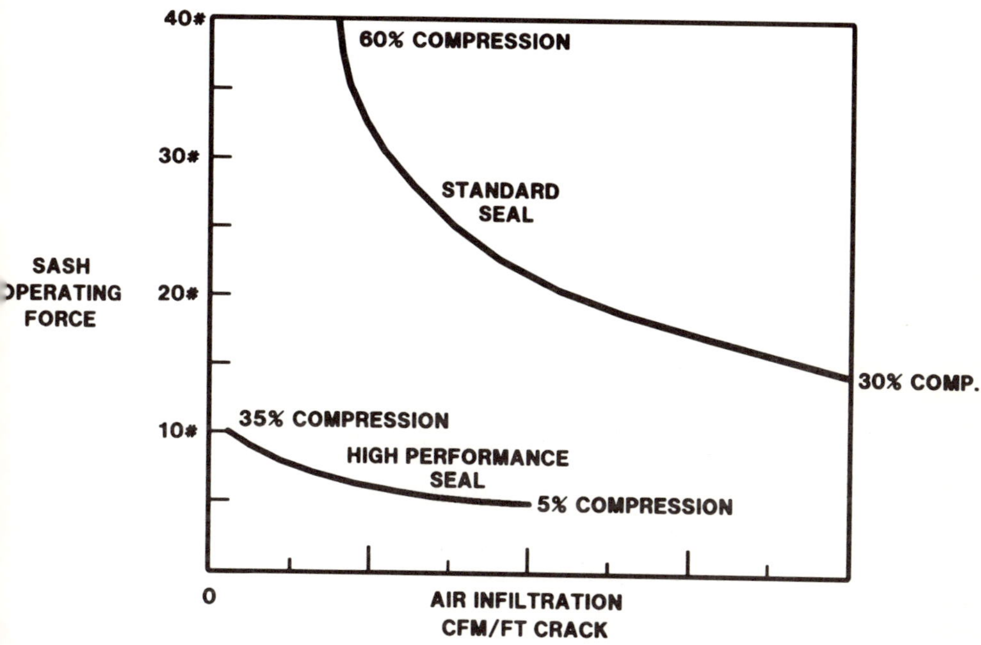

Figure 1. Air Infiltration vs. Operating Force.

The manufacturer's prime task in this area is to provide the lowest possible air leakage at a reasonable cost to the purchaser and with a window hardware design which provides long life and easy operation. If an examination of two types of pile weatherseal currently on the market made, results typical to those shown in Figure 1 will result. The graph plots air leakage against operating force for vertical and horizontal sliding sash which comprises approximately 85% of the windows used in today's housing. The comparison is of a standard pile as a continuous, impervious air seal. The high performance seal provides very low leakage while keeping the operating force well within acceptable levels. Numerical values of air leakage have not been given as pile construction and window design will affect actual numbers. However, the final numbers on any given window

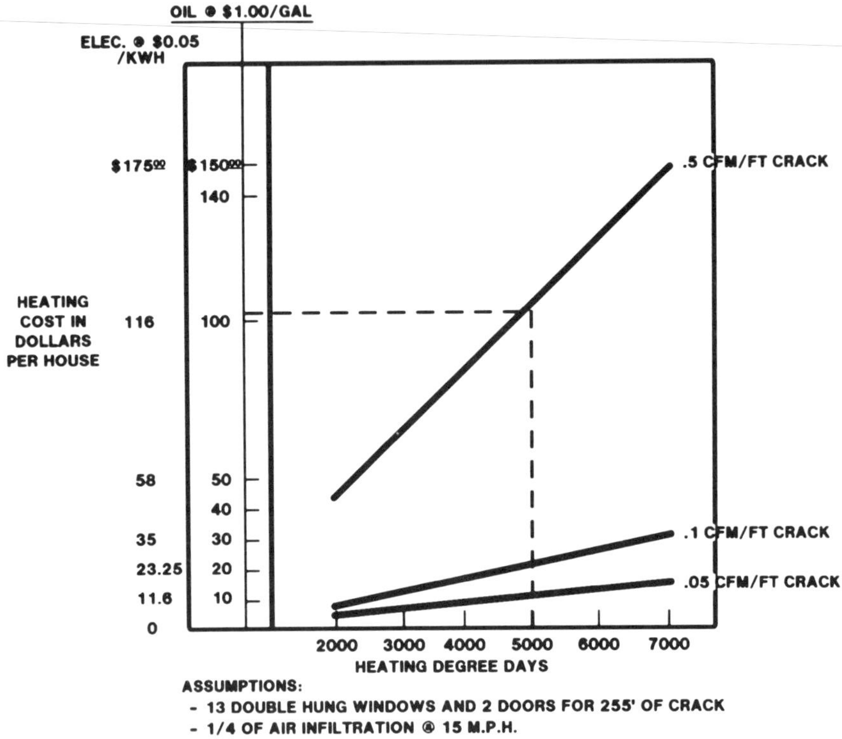

Figure 2. Cost of Air Infiltration.

design will follow these characteristic curves when tests are made on a window design using the two types of pile. For those who might wish to get specific values, at least one weatherstrip manufacturer offers a calculator which can be used to get fairly accurate results that have been proven in actual testing.[1]

The graph in Figure 2 shows the annual heating dollar cost of windows for a home at various levels of air leakage as a function of degree days. The conditions are: (1) thirteen double-hung 3X4 foot windows and two doors for a total of 255in. of crack perimeter, and (2) air infiltration at ¼ the air leakage rate when tested with a 15 MPH wind. This ¼ average rate is used since a heating season wind speed of 7½ MPH is more typical of much of the country and only ½ of the windows are leaking cold air at any given time.

At the industry standard of 0.5 cfm/ft of crack, the annual cost of heating the infiltrating air in a 5000 degree day climate for oil heat at $1.00 per gallon is approximately $100 per year for the assumed conditions. By using high performance weatherseals and good window design, this level can be reduced by a factor of five to under $20 per year. Conversely, the cost of air infiltration through poorly designed windows which are non-certified and do not meet industry standards can be two to three times the $100 per year value.

The next step a manufacturer can take to reduce air infiltration is to reduce the crack perimeter of a window. Single hung versus double hung and single slide versus double slide windows reduce leakage by virtue of design since one sash has been eliminated by glazing the glass directly in the frame. Thus, much of the crack perimeter in the window has been eliminated. By changing the conditions of the example in Figure 2 from 13 double hung windows to 13 single hungs, seasonal heating cost for air infiltration of the windows is reduced by 36% (use a multiplier of 0.64 times the dollar/year costs stated above).

As manufacturers, we have to assume that customers will pay a reasonable price in air leakage costs in order to achieve the natural ventilation benefits. The use of designs with high performance seals and the use of single vent sash does much to keep air infiltration at a minimum while still providing the benefits of natural ventilation.

Another performance compromise manufacturers must factor into their windows is that of solar gain versus conductive heat loss through the heating season. In coming years this will become increasingly important. Conductive heat loss takes place primarily through the glass and secondarily through the frame and sash. As long as the frame and sash are made of wood, aluminum with a thermal break, plastic, wood and aluminum, or plastic and aluminum, the heat loss through well designed windows with the same glazing arrangements will be about equal.

Most people are familiar with the materials mentioned but the term "thermal break" is often misunderstood. Many people believe thermal break to mean simply a window with sealed insulating glass. The trade name of "Thermo-Pane" has almost attained generic standing as synonymous with sealed insulating glass. A thermal break is more than that. It is a unit where the outer aluminum section is separated from the inner aluminum of the window frame by an insulating material, normally a plastic. In addition, a true thermal break window has at least two pieces of glass with an air space between.

Since sash and frame energy loss performances are approximately equal in well designed windows, the manufacturer's major concern are the design alternatives of using multiple layers of glass. The illustrations in Figure 3 show three different sash types which can make use of multiple glazing options. The first type with single sash can be either double or triple glazed with a sealed insulated glass element. The second example is a two sash

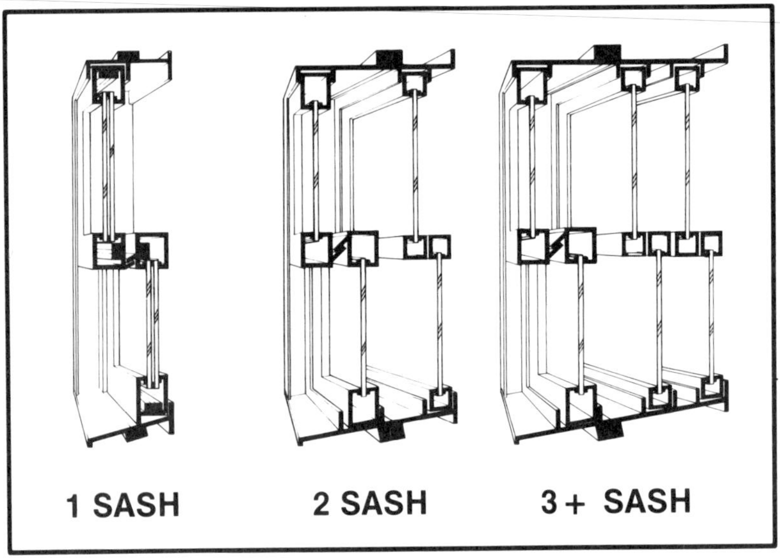

Figure 3. Multiple Glazing Designs.

window which is double glazed by using single glass panes in both sashes with a larger air space between the outside sash and inner sash. This unit can be triple glazed by the use of sealed insulating glass in either outer or inner sash, or can be quadruple-glazed by using sealed insulating glass in both sash. The last type is a three-plus-sash window. Each sash is single glazed. Three sash as shown gives a triple glazing—adding a fourth would give quadruple-glazing. The dead air space between each sash provides the insulating value.

Each of these window types with the multiple glazing options indicated offer the homeowner benefits but also drawbacks. Figure 4 lists seven factors to be considered in selecting the proper window type. The first of these factors is ease of cleaning. Since the single sash has only two surfaces which need cleaning, the rating is average. Double sash is poor because there are four surfaces to clean and the three sash window with six surfaces is rated very poor.

MULTIPLE GLAZING DESIGN EVALUATION

CLEANING		AVG. (2 SURFACES)	POOR (4 SURFACES)	VERY POOR (6+ SURFACES)
BREAKAGE	2G	POOR	AVG.	
REPLACEMENT	3G	VERY POOR	POOR	AVG.
WINDOW OPERATION		GOOD	AVG.	POOR
U-VALUE	2G	AVG.	GOOD	
	3G	POOR	GOOD	VERY GOOD
AIR INFILTRATION		AVG.	GOOD	GOOD
GLASS	2G	AVG.	EXCELLENT	
FAILURE	3G	VERY POOR	AVG.	EXCELLENT
SOLAR	2G	GOOD	GOOD	
GAIN	3G	AVG.	AVG.	AVG.

Figure 4. Multiple Glazing Design Evaluation.

In the area of breakage replacement, two sashed double-glazed and three sash units are given a rating of average because each sash employs a single pane of glass that should be easily replaced by the homeowner. The single sash with insulated glass and the two sash window using insulated glass are rated poor because special glass is needed for this product and availability of replacement glass is poor.

When window operation is considered, single sash is considered good because there is only one set of operating sash. The two sash unit is average because of the second set of sash, similar to storm windows added to normal windows. The three plus sash unit is rated poor because of the complexity of multiple sash operations.

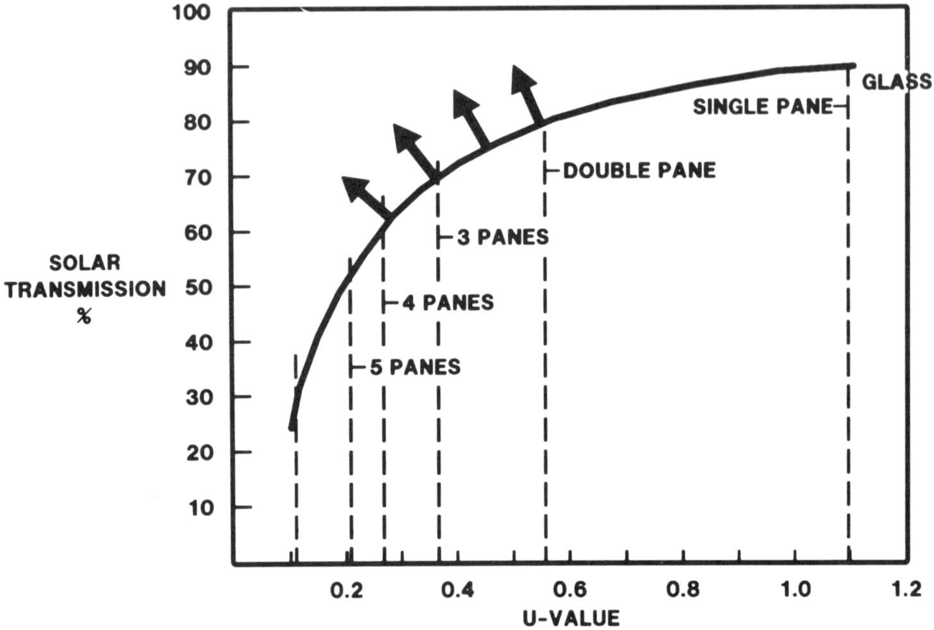

Figure 5. Solar Gain vs. U - Value.

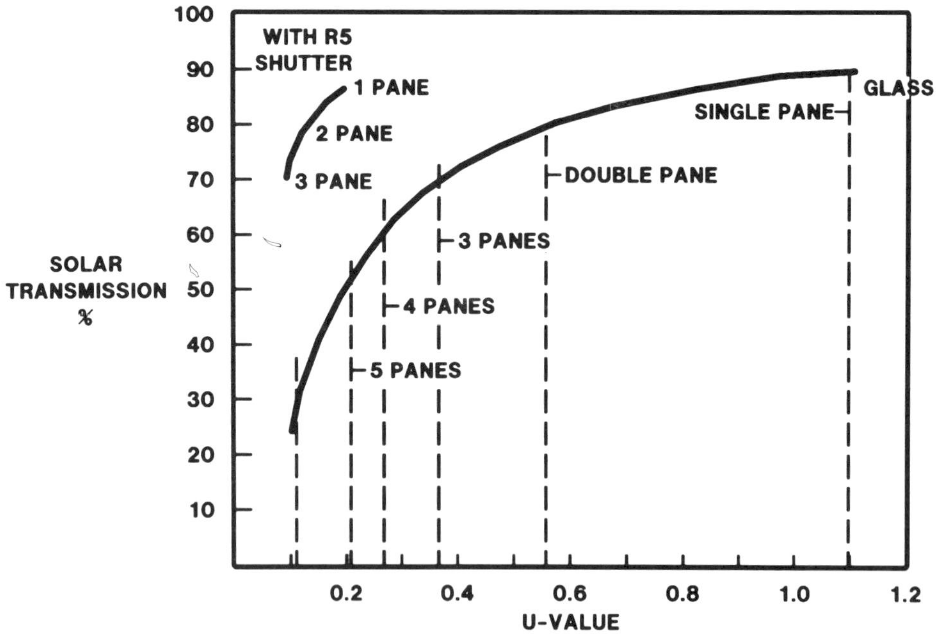

Figure 6. Solar Gain vs. U - Value.

The U-value rating assigned the single sash, dual insulated glazed window is average because there is some heat loss due to a thermal short circuit around the edges of the glass where a metal separator is located. The triple glazed single sash unit is rated poor because this type of triple glazing offers very little U-value improvement over a dual insulated glazed sash. The two sash window is rated good both for double and triple glass because of the wider air spaces and the thermal break in the metal frame significantly contributes to a lower U-value and prevents any thermal short circuiting between glass as the metal spacer of the insulating glass does. The three sash type window is rated very good because of multiple large air spaces between all glass and because thermal breaks prevent any heat loss short circuits.

For air infiltration, although the tightest set of sash normally determines the rate of air infiltration, under high wind velocity two and three sets of sash will diffuse the air with less direct single point leakage and greater comfort factor. Therefore, multiple sash get better ratings.

Glass failure of the insulating seal is not a consideration in the use of single glass so the single glass, single sash and the double glazed two sash unit get excellent ratings. The rating of units using double insulated glass is average because there is a possibility of seal failure. The single sash, triple insulated glazing is rated very poor because two seals are involved doubling the chance of failure. In addition, present manufacturers of triple insulating glass are currently not willing to provide a warranty for this product without high passed on cost to the consumer as protection against failure.

In the area of solar gain, the good ratings are given for those windows using two pieces of glass since a reasonable amount of solar heat still can pass through two pieces of clear glass. As the number of layers of glass is increased, the loss of solar transmittance becomes greater and a quad-glazed window would be a poorly rated unit from a solar heat gain standpoint.

An overall rating has not been assigned since other factors such as price, aesthetics, and structural location must also be considered. The ratings here are primarily from the viewpoint of each design used as a window. As the designs are viewed as passive solar collectors the primary factor to consider is the net BTU gain for a heating season. Figure 5 shows improvement of conductive U-value as more layers of glass are added to a window. However, as can be noted the solar transmission decreases as this U-value improvement is achieved. The area of the curve of major interest due to cost effectiveness as well as solar gain is between the two and four pane points. When more than three panes of glass are used, the drop in solar transmission is sufficiently high that such windows would probably not be considered for south facing windows acting as solar collectors. Quad-glazing for north faces may be cost effective, however.

One approach to achieving balance might be to consider a two sash window (see Figure 9) and leave the outer sash with double glass and place R5 sheathing material in the inside sash. Wallpaper or paint the inside to match the room and put a heat-absorbing coating on the exterior of the sheathing material. The use of the moveable shutter in the inside sash can then be

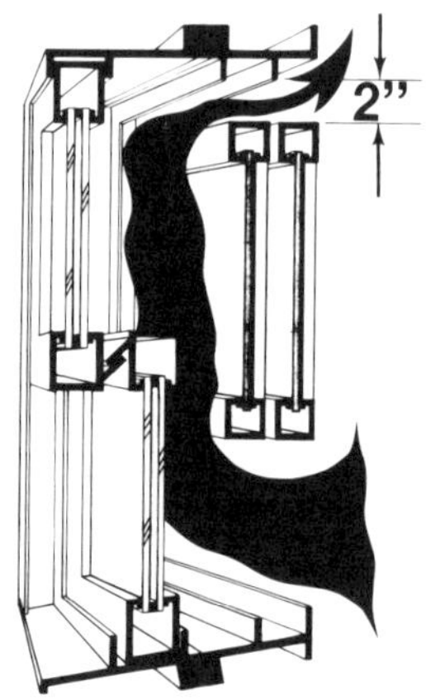

Figure 9. Passive Solar Window Concept.

managed easily by complete closing at night to reduce heat loss with an approximately 0.15 U-value barrier. The benefit of painting the outside of the shutter with a heat absorbing coating is realized in the daytime when the top sash is dropped two inches and the bottom sash is raised in line with it (Figure 10). In this position, convection currents carry the solar gain into the room as if the shutters were completely removed. The benefits to this type window are: a maximization of solar gain and U-value combination; reduction of summer heat gain; retention of tight sealing; the elimination of shutter storage problems. There is also a reduction of glare and fading (without loss of heat gain) and present windows can easily and inexpensively be retrofitted using new inside storm windows. Against these benefits are the fact that for these windows to be beneficial they must be managed twice a day. They may restrict daytime view if not used with high windows. They may also need a variance to meet existing HUD and FmHA codes such as the one for triple glazing in certain areas of the country.

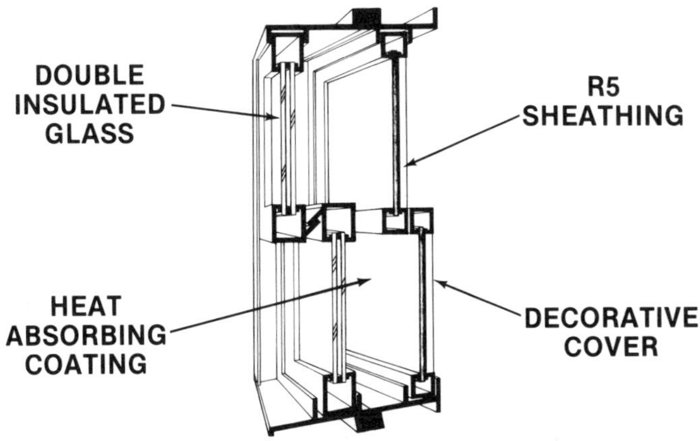

Figure 10. Passive Solar Window Concept.

Before you start considering using this window in the next house you are designing or building, please note that it is only at the concept stage at this point. But the idea demonstrates how a window manufacturer can respond to the growing market for windows as passive solar collectors. As a final point, consider the window manufacturer's perspectives. These include cost and return on capital employed. Before a manufacturer is going to invest a quarter of a million dollars (the current cost to design and develop a good new window), the manufacturer has to know that there is a market for the product. 99% of today's windows are still sold as windows, not passive solar collectors. We have economic incentives and the inconvenience of fuel shortages inspiring the marketplace, but there is a need for this product. Once the need is defined it may take up to a year-and-a-half to bring the window to commercial development-during which time other market changes as well as code requirements could impact on the product.

In addition, to gain the needed manufacturing space, the company will have to trim its existing lines. This is not always easy because changing codes and needs are taking windows nearing the end of the life cycle in northern climates and establishing them in a new life cycle in southern parts of the country right now.

The marketplace for residential and commercial construction is changing. The product requirements are changing. We are in the forefront of this change, particularly in work with solar energy.

The window manufacturers stand ready to assist during these changing times and come up with creative solutions to market needs. Five years from now, perhaps 99% of the products we make will be sold as passive solar collectors not windows.

The window industry has committed itself to researching and developing its product for energy efficient use and producing tools that may be used in the design and construction of homes. One example is the "Capitol Products" Window/Solar Energy calculator kit. This kit is currently available for $2.00 from Capitol Products Corporation, P.O. Box 69, Mechanicsburg, PA 17055. It can be used by homebuilders and designers to design homes using windows as passive collectors. Not only does the calculator consider different geographic areas and the effects of window management, but it permits the calculation of overall cost savings for different fuel types. The information in this slide rule type calculator was developed from a computer program using actual hour by hour weather tapes. The data was compiled by the Hittman Associates for the funding body-the AAMA Research Foundation.

In the same vein, a weatherstrip design calculator is available from the Schelegel Corporation, 1555 Jefferson Road, Rochester, NY 14601. Write for it attention Fred Williams.

Natural Air Movement Through Buildings

Benjamin H. Evans, AIA

Professor, College of Architecture and Urban Studies
Virginia Polytechnic Institute and State University
Blacksburg, Virginia

The subject of natural ventilation in buildings seems to be a strange one for most of us. We subconsciously know that the wind blows outdoors and that it can be pleasant or unpleasant. We also know that it can be useful indoors. But we seldom give much conscious thought to making use of natural air movements and one of the reasons is that there is very little substantive information available on the subject so that we all tend to be rather uninformed. Another reason is that air movement is quite difficult to predict and therefore difficult to design for. This paper will point out a few of the basic principles involved and suggest some approaches to design that you may find useful.

Natural ventilation has three principal aspects for our concern: (1) it can be used for cooling or warming *people*; (2) if can be used for cooling or warming *objects*; and (3) it can be used to provide *fresh air* for breathing. Since the fresh air aspect is normally only a problem if we are dealing with closed mechanical systems, it will not be considered further.

Human comfort is, as we all know, affected by air temperature, humidity, air movement, and radiation. Air movement and radiation are the factors least used and probably most misunderstood.

Consider using natural air flow for cooling. Air movements produce a cooling tendency in people by accelerating the conduction of heat from the body and by increasing the opportunity for evaporation of skin moisture. During warm weather, air movements can be used quite effectively in this fashion. There is hardly a community in the United States where cooling with natural air flow in warm weather is not possible and profitable for at least part of the year. But, even when the air is moving where it can carry off body heat, there are some characteristic constraints. For instance, some people have minimal skin moisture (they don't perspire much). People wear clothing (even in warm weather) which restrains air movement from doing its evaporative job. Some people do not like other effects of the wind. The bookkeeper with the piles of loose paper might not like to have the papers blown about. In extremely dry climates where body moisture will evaporate easily, direct wind currents are not only unnecessary but undesirable.

Therefore, design for natural air movement is not the cure to all that ails the world. In Houston, Texas, when the air temperature is 95 degrees and the humidity is 65 percent, almost everyone will want to retreat to the nearest air conditioned building, regardless of how the winds blow. Still, that leaves ample opportunity for the use of natural air flow in wide parts of the country and in a variety of seasons. The trouble is, people have simply gotten out of the habit of using air movement effectively. Our buildings are not oriented properly, they are not designed for "cross ventilation", and if they have operable windows, they will probably not move the air for best effect.

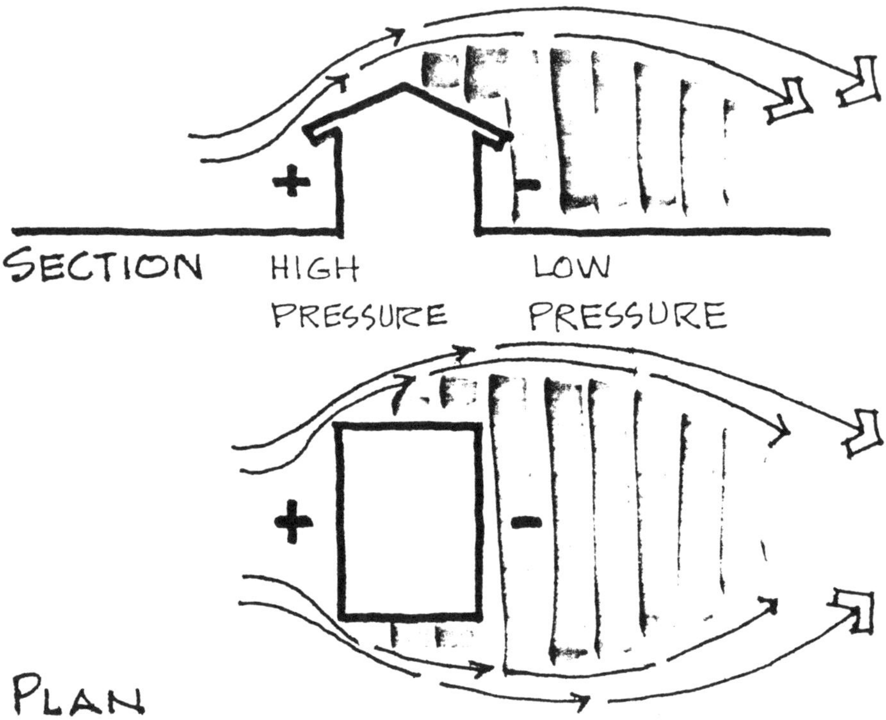

Figure 1. Pressure Differences.

Design for air movement cannot, of course, be considered in isolation from the other important concerns in building design. This paper will concentrate on air flow and leave it to other designers to determine how all the conflicting parameters are to be resolved.

There are several basic characteristics of air movement:

(1) Air is moved by *pressure differences* and it moves from high to low pressure.

(2) Air has *bouyancy*. As it expands with increasing temperature, it becomes less dense and tends to rise above a more dense cooler air mass.

(3) Air has *inertia* and, once set in motion, tends to continue its direction of movement until some external force causes it to change its direction.

(4) Air movement tends to lose some of its inertia because of *friction* between the air and adjacent surfaces as it moves along.

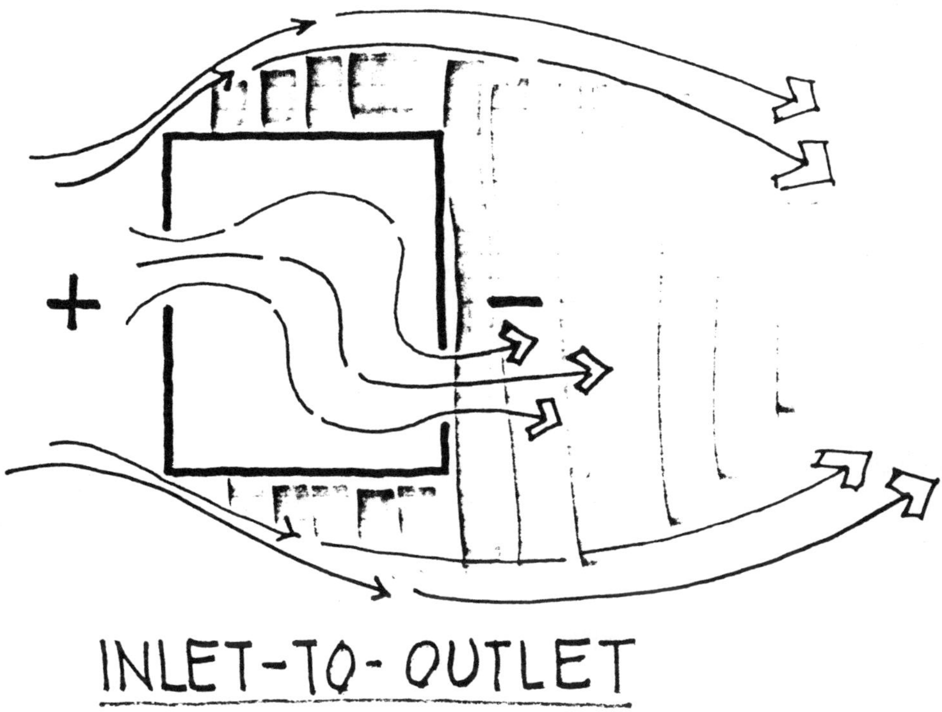

Figure 2. Pressure Differences.

Of fundamental importance is the consideration of where pressure differences are created when the wind impinges against a building.[1] A region of high pressure is created, generally on the windward surface of the building. As the air is deflected around and over the building, it speeds up and causes relatively low pressures along the sides just behind the windward face and along the lee side.

Naturally, air tends to flow into the building at the high pressure points (windows, cracks, etc.) and out at the low pressure points. The speed at which the air moves through the building is determined by the relative magnitude of the pressure areas created by the impingment of oncoming air and by the location and size of window openings—inlets and outlets.

The greatest differentiation of pressures will be created when the air movement is perpendicular to the windward face of the building. In circumstances where the building is made up of combinations of shapes, or odd shapes, or faces the wind at odd angles, the prediction of pressure differentials can be very difficult. In fact, except for the simplest situations, the only way to determine pressure differentials and resulting air flow patterns is through the use of wind tunnel studies of scale models.[2]

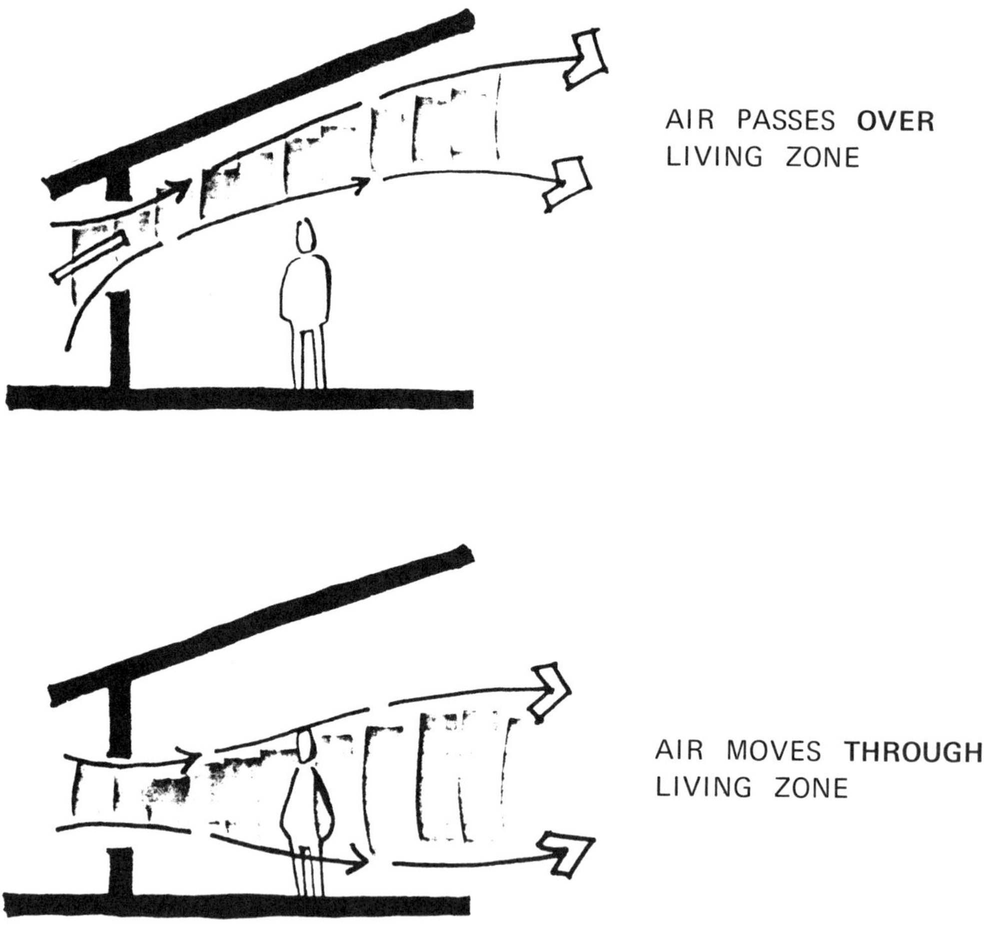

Figure 3. Location of Ventilator Opening Affects Air Flow.

But, just moving the air through windows and into (and out of) the building is not all there is to producing effective cooling of the occupants inside.[3] The air must move through the "living zone" of the space where the occupants will feel the movement of the air. There is some value to moving air across the ceiling, for instance, to carry off the hot air build-up, but that air moving above the occupants will not be very effective in causing a cooling sensation. Properly utilized windows and inlets are among the most effective elements for giving the incoming air the proper directional component.

There are three main factors which determine the air pattern as it enters the building. These are: (1) the location of the inlets in the wall; (2) the type and configuration of inlets used; and (3) the location of other objects adjacent to the inlet such as landscaping, overhangs, and louvers.

Consider the location of the inlet in the wall. As the air impinges upon the face of the building, it builds up velocity vectors which move across the surface. If the inlet opening is located in the center of the wall, these vector currents will tend to be symmetrical and the air will enter the inlet in a "straight in" pattern. If the inlet opening is located off-center, as in the third story wall shown in this illustration, the air will have a tendency,

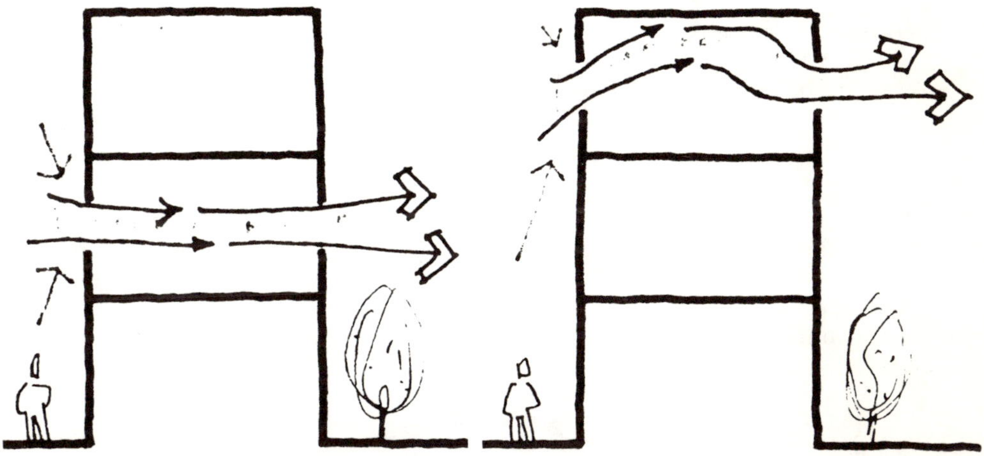

FORCE VECTORS ABOUT EQUAL FORCE VECTORS UNEQUAL

Figure 4. Cross Section Showing Effect of Inlet Location.

because of the surface vector forces, to enter the building in an upward direction. Or if the opening is in the first floor, the air will have a tendency to enter the opening in a downward direction, although the ground surface also has its effect on the building surface vectors. These variations in air patterns are caused by the inequality of pressure above and below the inlet openings as a result of their location in regards to the proportion of solid wall surface around them.

The same principle applies to the air patterns resulting from the location of the inlets as viewed in plan.

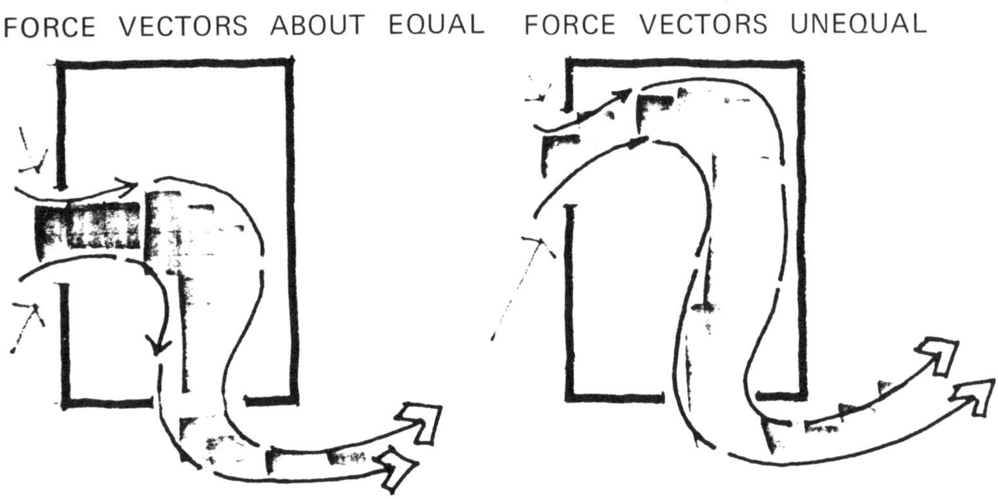

Figure 5. Plan View Showing Effect of Inlet Location.

The outlet is irrelevant to the pattern of the air entering the building. The inertia of the air entering the inlet opening causes it to flow upward and across the ceiling regardless of the location of the outlet opening, and continue across the ceiling until its path is obstructed and it must turn to get to the outlet.

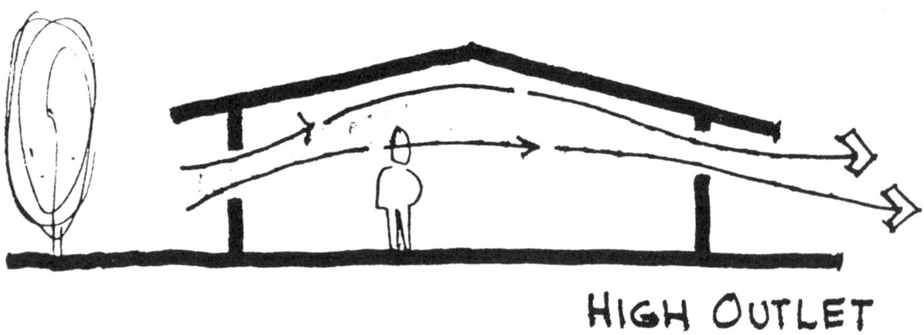

Figure 6a. A High Outlet Tends to Direct Movement to the Ceiling.

Figure 6b. A Low Outlet Keeps Movement of Air in the Living Zone.

The configuration of the inlet itself is a major determinate of the incoming air flow patterns. This illustration shows what happens when a vertical projection such as a door, shutter, or a casement window is added to a simple opening. The force vector (or component) from the larger wall surface area is disrupted, allowing the force vector on the smaller wall surface area to become dominant and the pattern of the entering air is altered dramatically.

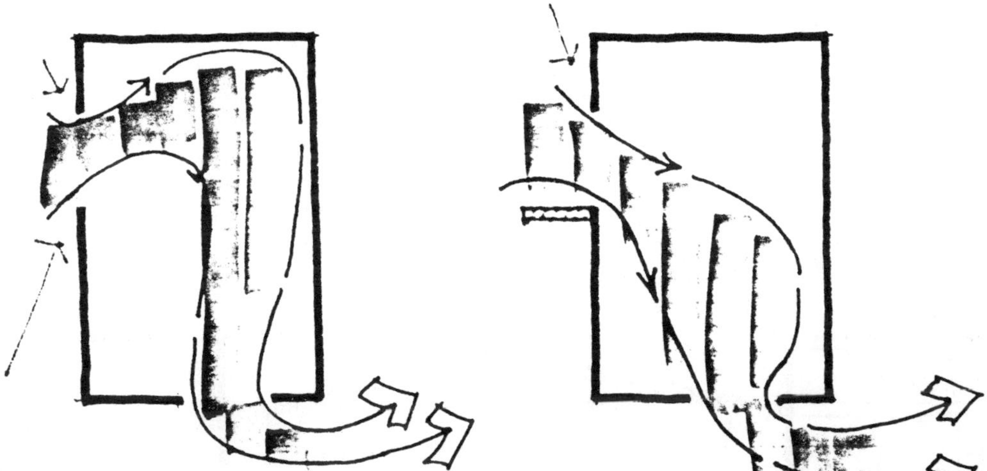

Figure 7. Placement of Projections Modify the Force Vectors and Resultant Air Flow Pattern.

A similar effect can be seen when a horizontal projection is added over a window. One might say that the inlet configuration acts very much as the nozzle of a hose and gives direction to the air passing through.

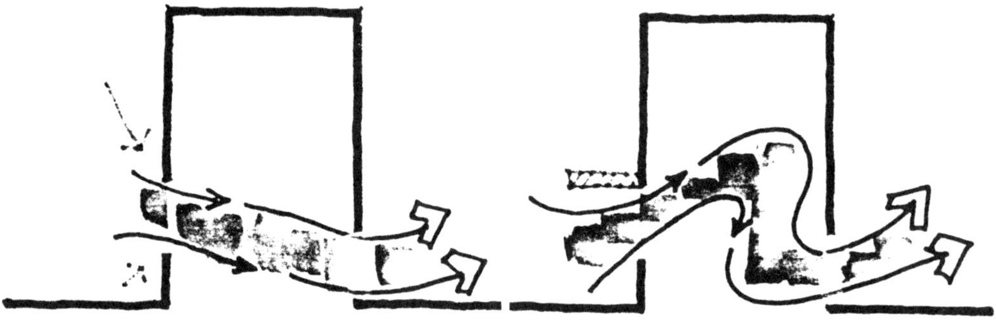

Figure 8. Horizontal Projections over or under a Window Modify Force Vectors and Air Flow.

While there are a great variety of windows on the market, they may be classified as one (or a combination) of three basic types as related to air movement:

(1) **Simple opening** — windows such as the single or double hung and horizontal sliding—any window which opens by sliding in a single plane.

(2) **Vertical Vane** — windows such as the side hinged casement, the folding casement, and the vertical pivoted or reversable window—any window which opens by pivoting on a vertical axis.

(3) **Horizontal vane** — windows such as the projected sash, the awning, the basement, the horizontal pivoted, and the jalousie—any window which opens by pivoting on a horizontal axis.[4]

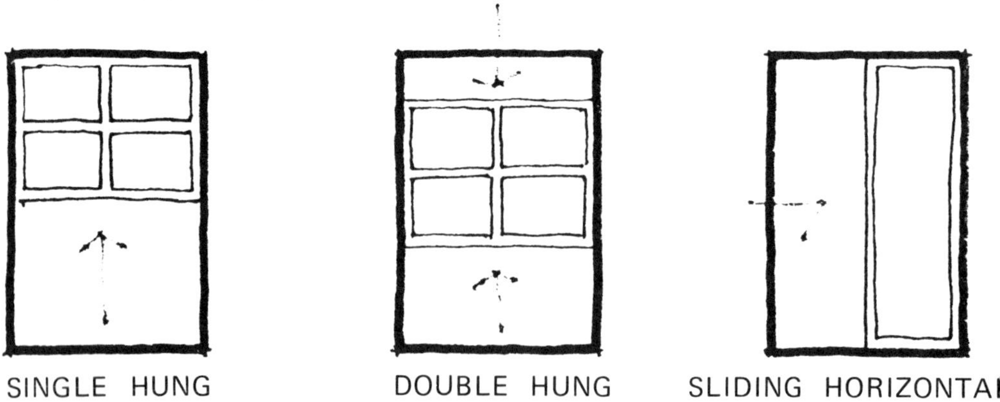

Figure 9. Sliding Type Window Openings.

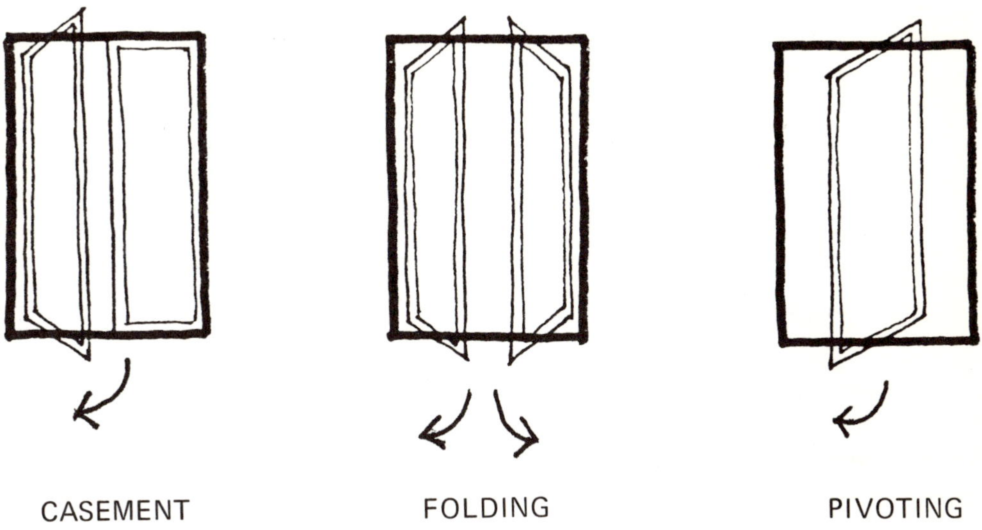

Figure 10. Vertically Hinged Window Opening Types.

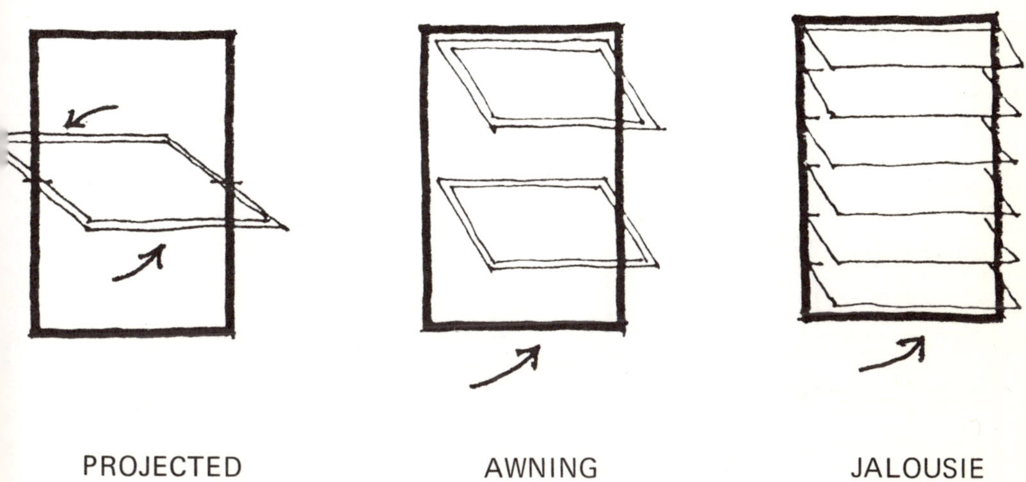

Figure 11. Horizontally Hinged Window Opening Types.

Air patterns resulting from some typical inlet types are shown in section and plan in these illustrations.

Air movement through simple window types tends to be governed by the exterior wall surface conditions or force vectors.

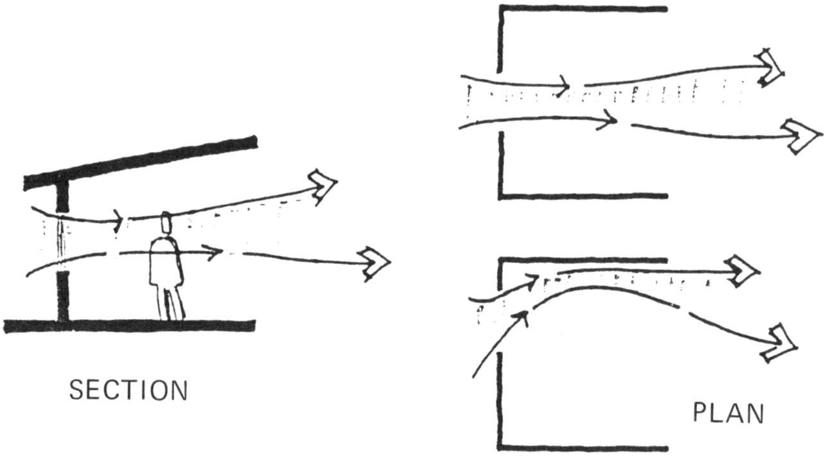

Figure 12. Sliding (Simple Opening) Type Flow.

With vertical vane windows, the air is directed by the vanes (in plan view) and generally flows along the surface of the adjusted window sash (vane), just as the flow of air follows the rudder on an airplane's tail. The direction of the air in section is still controlled by the force vectors of the exterior wall surfaces above and below the window.

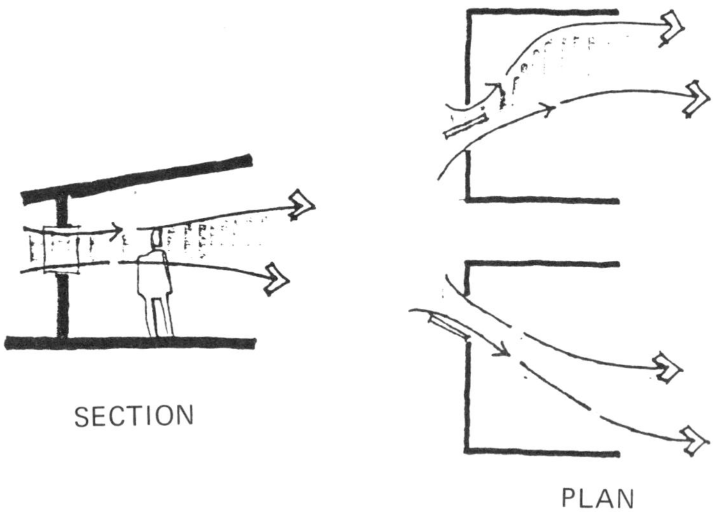

Figure 13. Vertical Vane Type Flow

The horizontal vane window works on the same principle as the vertical vane except that the operating sash directs the air up or down rather than sideways.

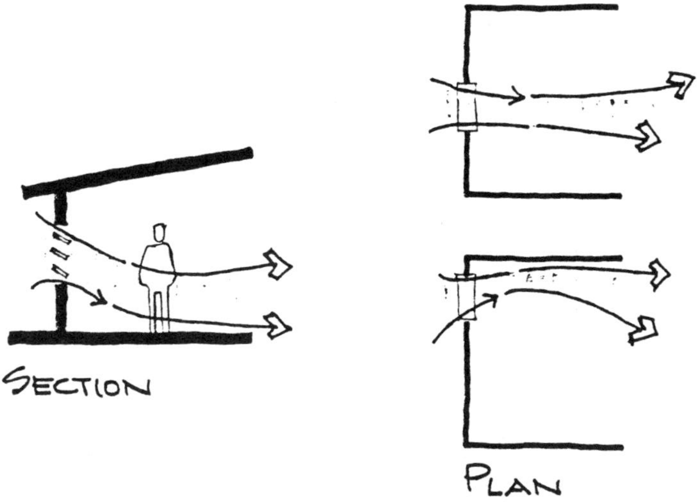

Figure 14. Horizontal Vane Type Flow.

Of course, windows are not the only medium for moving air through a building. Sometimes, there are valid reasons for keeping the "window" separate from the operable ventilating component.

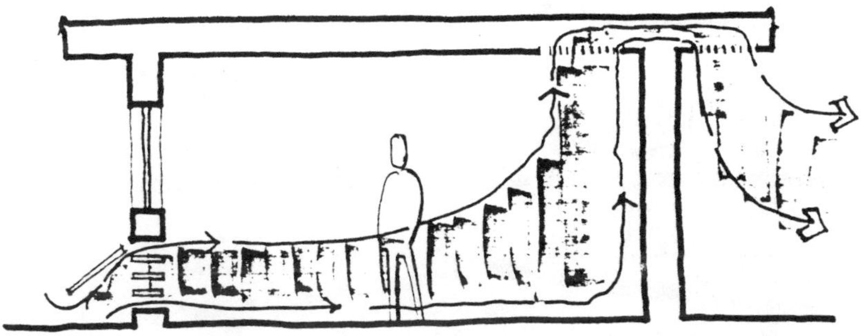

Figure 15. "Non-Window" Openings.

As noted previously, the speed with which the air moves through the building is principally a function of the relative air pressures at the inlet and outlet and the sizes of these openings. Increased air speeds within a building are achieved when the outlet is larger than the inlet. Inertia and friction will play their part to some degree in slowing air movement but their effect is relatively minor.

Theoretically, the greatest volume of air flow, or the greatest number of air changes, is achieved when the outlet is, at least, equal in size to the inlet. To maximize total volume, both the inlet and the outlet should be as large as possible.

But this volume of air change view of natural air flow has little to do with summer cooling. A building may have many changes per hour, but if the air does not flow through the living zone and if the air speeds are slow, the occupants will not experience a cooling sensation.

Many people incorrectly assume that good air flow is achieved when the large openings are faced toward the breeze with the idea of "scooping" the air into the room, along with some small openings on the opposite side of the room to allow "cross ventilation". Actually, the reverse situation would be better from the standpoint of summer cooling, assuming the inlets are properly designed and placed to direct the air into the living zone.

This figure shows air speeds through typical building sections as tested in a wind tunnel.[4] The oncoming air speed is indicated as 100 and the inside air speeds are expressed as a percentage of this outside speed. The diagrams clearly indicate a substantial gain in inside air speeds as the size of the outlet is increased while holding the inlet size constant.

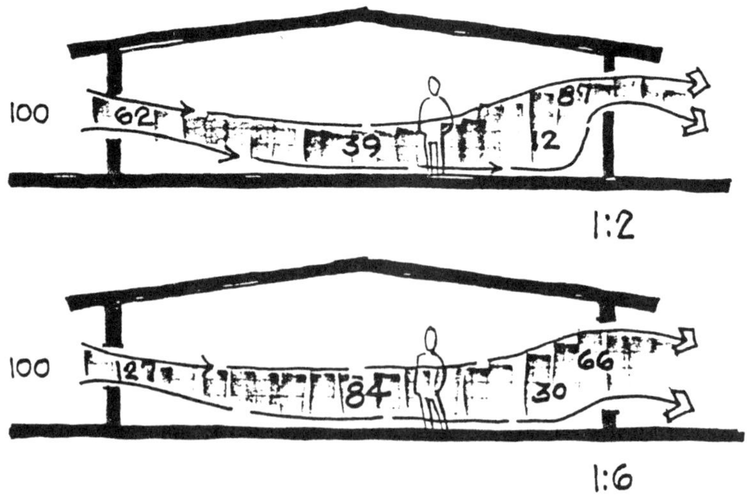

Figure 16. Effect of Inlet/Outlet Ratio.

This figure shows the effect of the location of the outlet opening on air speed. Because of the air's intertia, when it is forced to change its direction of movement its speed is retarded somewhat and potential summer cooling energy is wasted. Changes in the direction of the air movement through a building should be kept to a minimum.

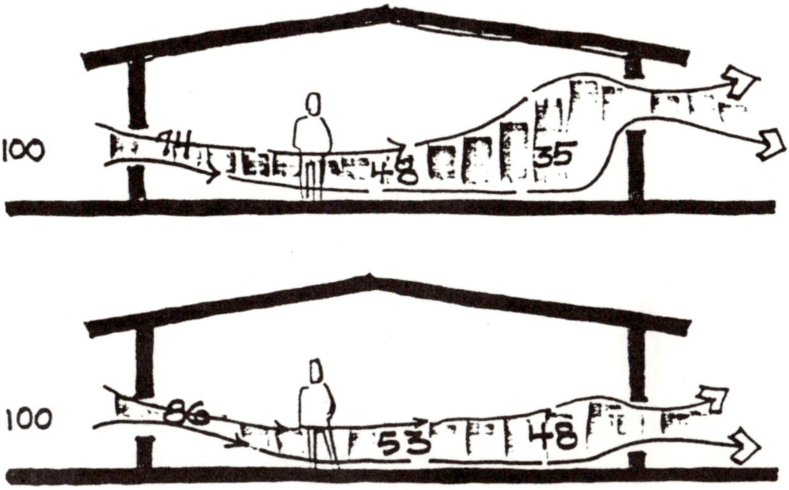

Figure 17. Turns in Flow Cause Friction.

During the post-World War II school building boom, before air conditioning became commonplace, a number of schools were built in the southwest with natural air flow as a major design determinant by architects Caudill, Rowlett, Scott. I was their air flow consultant. These figures illustrate two such designs.

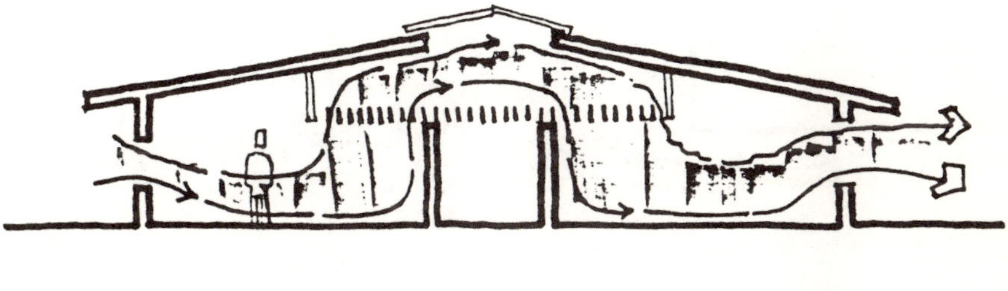

Figure 18. Miami, Oklahoma.

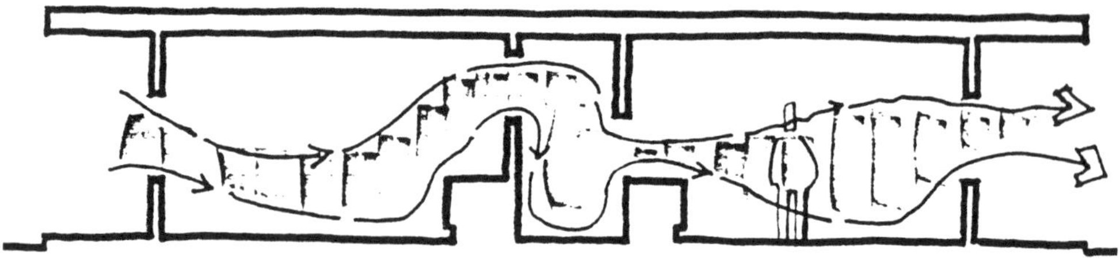

Figure 19. Laredo, Texas.

These figures show some of the effects of landscaping on the air flow through a building.

Figure 20 shows how high hedges may be used to create pressure differences and effectively cause air to flow through a building oriented with no inlet openings facing into the wind and which would otherwise be without significant air movement. This suggests that buildings do not necessarily have to be oriented perfectly to the wind to make good use of the cooling effect of moving air.

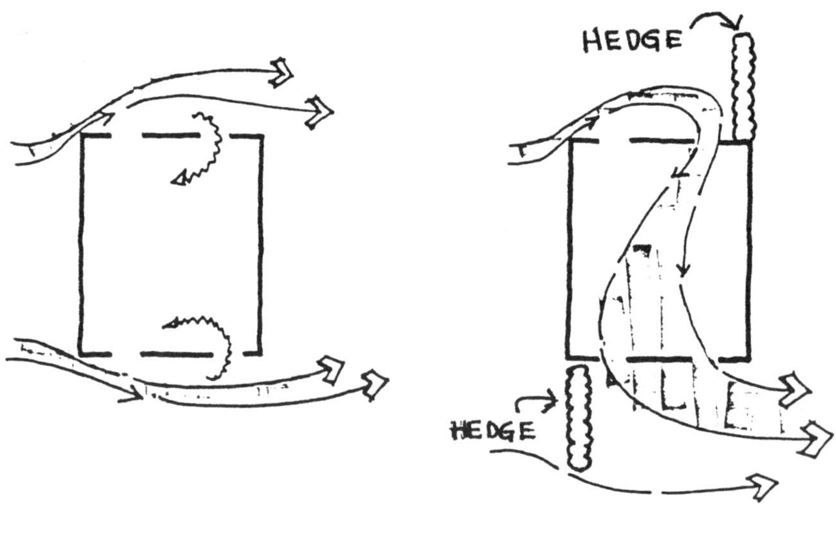

Figure 20. Hedges Can Be Used To Direct Air.

This figure shows the effect of variations in a hedge and tree arrangement. With the hedge located leeward of the tree as shown in the first diagram, the air flow is deflected over the building creating a low pressure region in front which causes the air to be drawn through the building slowly in the opposite direction to the wind. But with the hedge and tree arrangement reversed as shown in the lower diagram, the air flow is directed back to the ground and through the building where its cooling effects can be felt.

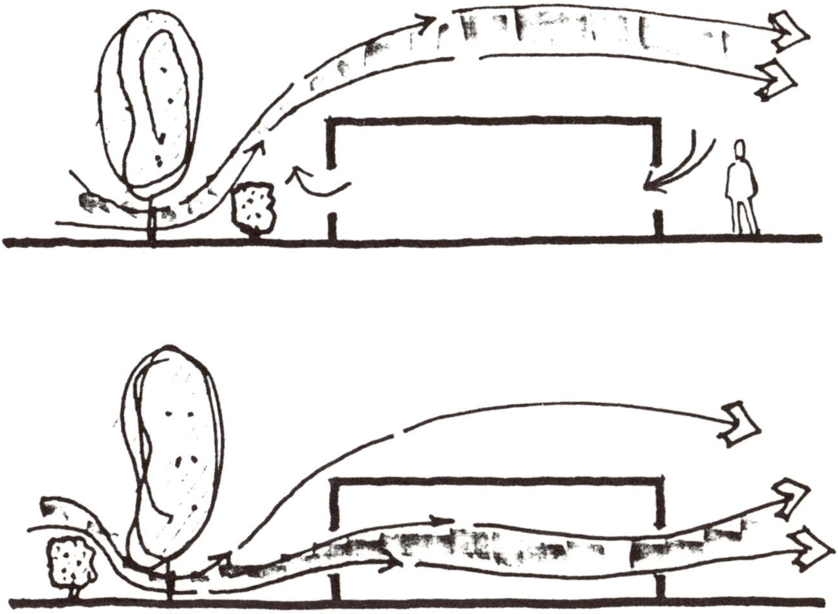

Figure 21. The Effect of Tree and Hedge Combinations.

Screens represent another effective tool for controlling the flow of air. The solid screen presents a substantial barrier that must be bypassed by the air.

As the air strikes the screen, it is deflected up and over, thus causing a high pressure area upwind and a low pressure downwind and an eddy area behind the screen. The speed of this eddy current can be quite significant relative to the speed of the air movement upwind—in some cases as much as 50% of the upwind air speed.

If a perforated screen with a 30% to 50% opening is used instead of the solid screen, the speed of the eddy will be less than for a solid screen, except for the area immediately adjacent to the screen. The air passing through the small perforations in the screen effectively slows the eddy pattern so that the perforated screen actually offers more protection from winds than the solid screen.

Another area of interest is air movement by convective currents. It is one of the passive designer's most cherished options and probably one of the most misused. In a closed circuit system (i.e., a pressure tight loop), if the air is heated at one point, that warm air will tend to rise and the cooler air will tend to fall. But there is very little real data available to enable us to put some quantitative figures to the speed with which the air moves, or what volume will be moved, or how much heat will be transferred. And in an open-plan building, even if the design is considered a closed circuit system, it is even more difficult to accurately determine where those

Figure 22. Convective Air Flows due to Gravity.

movements will occur. Be wary of those thermal current diagrams you see so beautifully rendered in some designer's sketches. They may or may not work. Air will sometimes flow through a structure by temperature differences and the air can be usefully channeled in cold weather as an aid in heat distribution, but during warm weather this movement is virtually useless for cooling as compared to the cooling effect of even a very light breeze resulting from pressure differences around the structure. Studies in Australia have shown that in industrial plants with significant process heat from furnaces, even slight breezes will overcome the convective currents.[6] So, if you are going to design to try to make use of convective currents, be sure you are working with a closed circuit system.

In climates where there is a significant difference between day and nightime air temperatures (a large portion of the U.S.), nightime air can be brought into a building, not only for cooling people but for cooling the building also. Daytime heat stored in the building's mass can be exhausted to the exterior so that on the succeeding day radiant convective heat from those building masses will not contribute so greatly to discomfort. The approach to nightime mass-cooling is opposite that of daytime body-cooling in that usually much larger quantities of air volume must be moved through the building interior and speed of the air flow is otherwise not important. Nightime air humidity can be a deterring factor if the air is too moist, but for non-air conditioned buildings such as residences, apartments, schools, and light commercial stores, nightime cooling makes sense.

In summary, natural air flow can be an effective tool for energy conservation for the improvement of thermal comfort and for preserving some of the joys of nature, when its characteristics are understood and buildings are designed with proper attention paid to human interaction.

REFERENCES

1. Evans, B. H., "Natural Air Flow Around Buildings". *Research Report No. 59*, Texas Engineering Experiment Station, College Station, Texas, March, 1959.

2. Smith, E. G., "The Feasibility of Using Models for Predetermining Natural Ventilation", *Research Report No. 26*, Texas Engineering Experiment Station, College Station, Texas, June, 1951.

3. Caudill, W. W., Crites, S. E., and Smith, E. G., "Some General Considerations in the Natural Ventilation of Buildings", *Research Report No. 22*, Texas Engineering Experiment Station, College Station, Texas, February, 1951.

4. Holleman, T. R., "Air Flow Through Conventional Window Openings", *Research Report No. 33*, Texas Engineering Experiment Station, College Station, Texas, November, 1951.

5. White, R. F., "Effects of Landscape Development on Natural Ventilation of Buildings and their Adjacent Areas", *Research Report No. 45* Texas Engineering Experiment Station, College Station, Texas, March 1954.

6. *The Natural Ventilation of Industrial Buildings*, Department of Labour and National Service, Commonwealth of Australia, 1948.

Glass and its Role in Sun Control and Thermal Transfer

John I. Yellott, P.E.
Professor Emeritus
College of Architecture
Arizona State University
Tempe, Arizona 85281

INTRODUCTION

The following paper has been prepared as a sequel to the writer's presentation at the Third National Passive Solar Conference in San Jose, California. That paper presented the ASHRAE procedures for calculating solar heat gains and conduction heat gains and losses through unshaded single and double glazing. The present paper discusses additional thermal performance characteristics of glazing and reviews some of the non-solar design issues such as physical properties of glazing which are frequently overlooked. The writer is indebted to ASHRAE Technical Committee 4.5, Fenestration, and the *1977 ASHRAE Handbook of Fundamentals* for much of the material that is presented here.

GLASS AND ITS PROPERTIES

Architectural glass is a remarkable material which has been used effectively in buildings for well over a thousand years. The properties which must be considered in selecting glazing include: (a) physical properties such as strength, moisture permeability, thermal expansion and weight per unit area; (b) optical properties including transmittance, reflectance and absorptance; (c) thermal properties including heat conduction and emittance for longwave radiation; (d) properties determining longevity, including resistance to deterioration due to ultraviolet radiation, oxidation, staining, etc. These properties have been studied in great detail by glass manufacturers in the U. S. and abroad and reliable information is available in their literature and in references such as the *ASHRAE Handbook of Fundamentals.*[1]

PHYSICAL PROPERTIES

Considering first the physical properties, the soda-lime float glass generally used for architectural purposes has a density of about 0.09 lb/in^3. Of more interest to the designer of windows is the weight per unit area for the various thicknesses which are available commercially, as shown in Table 1 below:

Nominal Thickness, in.	3/32*	1/8†	3/16	1/4	3/8	1/2
mm.	2.5	3	5	6	10	12
Weight, lb/sq ft	1.23	1.64	2.45	3.27	4.90	6.54
Maximum cut size, sq ft or width x height, in.	25	35	70	128 x204	124 x260	120 x260

* 3/32 inch glass is often called "single strength"
† 1/8 inch glass is often designated as "double strength"

Table 1. Weights per Sq. ft. and Maximum Available Sizes of Architectural Glass

The use of double or triple glazing simply multiplies the unit weight by a factor of two or three.

Glass is very strong in compression but it is relatively weak in tension and, because it is a very brittle material, its performance is difficult to predict. In the glass industry, extensive tests have been run to determine the relationship of wind loading to breaking stress and it has been shown that breakage is related to process fabrication, surface quality, support conditions and manner of loading. Ordinary architectural glass is "annealed" and the wind loading which it may be expected to sustain is shown in Fig. 1 as a function of wind velocity and glass area. Single strength glass is used only in very small windows and it is not shown on Fig. 1. Where wind speeds above 60 mph never occur, double strength (1/8 in.) glass may be used for areas up to 40 ft^2 but for larger areas or where higher wind speeds are probable, ¼ in. glass should be used.

Maps of the U. S., Fig. 2, are available[2] which show the "Fastest Mile Velocities" for an elevation of 30 ft above local grade. In coastal regions along the Atlantic Ocean and the Gulf of Mexico, 90 to 120 mph must be expected at the 30 ft elevation. Along the Pacific coast, comparable values range from 67 mph in Southern California to 95 mph in Washington and Oregon. 75 to 100 mph winds are recorded for much of the interior of the U. S. Local building regulations should be consulted when large glazed areas are being designed. Lower velocities than those cited above generally prevail at ground level. Conversely, wind speed increases with increasing height above the ground.

It must not be forgotten that the large windows which are often recommended for use on the south exposure of passive solar houses are likely to use ¼ in. thick glass rather than the half as thick "double strength", and, as section 1.2 will show, this can have significant impact on the admission of both heat and light.

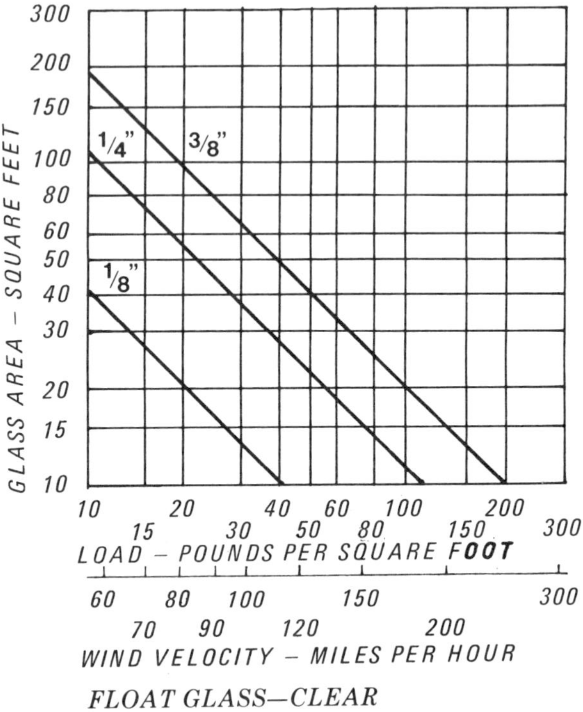

Figure 1. Wind Load Data for Architectural Glass.

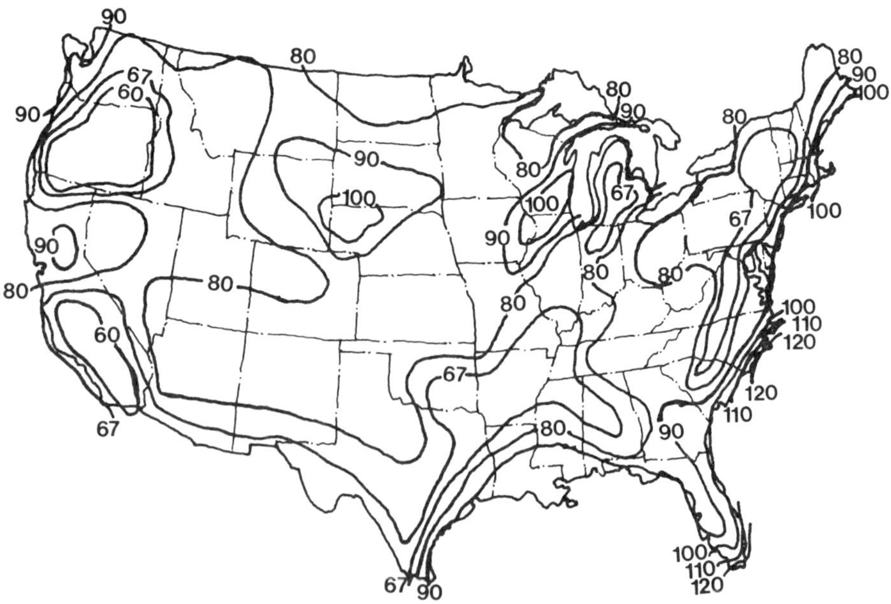

Figure 2. "Fastest Mile" Wind Velocities in the Continental U.S.

Very large increases in strength and thus reduced danger of breakage can be produced by heat treating, and glass which is fully tempered is considered to be nearly five times as strong as ordinary annealed glass. Another advantage enjoyed by tempered glass is its propensity for shattering into a myriad of tiny and relatively harmless fragments on the rare occasions when it does break. Tempered glass is required by safety codes in many applications and, since its cost is not much greater than that of ordinary annealed glass, it should be used whenever there is any question about the safety of a glazing installation. Glass breakage is still a poorly understood phenomenon but it is recognized that edge imperfections, such as nicks, notches or other cutting flaws, often become the starting points for cracks. For further details see Ref. 1, Chapter 26, or the literature of the leading glass manufacturers.[2]

Thermal expansion is a physical property of glass which is often overlooked but it must not be neglected in fenestration design, because, like most other solid materials, glass will inevitably expand when it is heated. The coefficient of thermal expansion, defined as the rate of expansion in inches per inch of length per degree F change in temperature, is given in Table 2 for glass and a number of materials with which glass is often associated. Glass must always be installed with adequate provision for thermal expansion and particular care must be taken when clear glass, which is generally relatively cool, is converted into heat-absorbing or reflective glass by the addition of adhesive sun control films. Manufacturers' instructions should be followed when large lights of glass are being installed. An excellent motto is: "Handle with Care!"

Aluminum:	12.8	Glass, plate:	4.5	Wood, parallel to the grain:	
Bronze:	10.1	Glass, pyrex:	1.8	Pine, oak:	3.0
Copper:	9.3	Brick Masonry:	3.1	Plastic:	36 to 52
Steel:	6.7	Concrete:	7.9	(methyl methacrylate)	
				Plastic materials often have very high coefficients. consult manufacturers for details.	

Table 2. Coefficients of Thermal Expansion for Building Materials (Multiply values by 10^{-6}).
(adapted from references 2 and 3)

For all practical purposes, glass is impervious to moisture and it is an excellent electrical insulator. Windows, on the other hand, frequently have cracks through which both moisture and air can enter when the air pressure outdoors is higher than that indoors. Capillary action can cause water to enter even when it must apparently flow uphill to do so, so it is important to remember that "glass can't leak but windows can!"

OPTICAL PROPERTIES

"Optics" is generally defined as the study of light but in this discussion the definition is broadened to include the performance of glass over a much broader range of electromagnetic wavelengths than the relatively narrow range of 0.4 to 0.75 micrometers which excites the sensation of vision in the human eye. Of particular concern are the so-called "solar-optical properties" which include transmittance, reflectance and absorptance as they apply to the shortwave radiation (0.3 to 3.0 micrometers) from the sun and to the longer infrared radiation (5.0 to 50 micrometers) which is emitted by sun-warmed surfaces. The latter may be dismissed quite readily by the simple but accurate statement that ordinary architectural glass is entirely opaque to longwave radiation and all of the infrared radiation beyond 5 micrometers that strikes a glazed surface is either reflected or absorbed. For the shortwave solar radiation, the situation is quite different.

Transmittance is the most important of the solar-optical properties and it depends upon: (a) the chemical composition of the glass and its thickness; (b) the angle of incidence between the incoming solar rays and a line normal (perpendicular) to the surface of the glass; (c) the wavelength of the incoming radiation.

TYPE AND THICKNESS

	Single 1/8″	Single 1/4″	Double 1/8″	Double 1/4″	Triple 1/8″ (est.)
Manufacturer A:					
Visible Trans.	0.90	0.89	0.81	0.77	0.72
Total Solar	0.85	0.78	0.69	0.59	0.51
Manufacturer B:					
Visible Trans.	0.89	0.87	0.80	0.77	0.71
Total Solar	0.80	0.75	0.69	0.56	0.51

Table 3. Visible and Total Transmittance for Architectural Glass.

Transmittance can be measured in two ways. The first, generally employed by glass manufacturers to rate their products, uses a spectrophotometer to measure the near-normal spectral transmittance at wavelengths from 0.3 to 3.0 micrometers. The transmittance at selected wavelengths is then multiplied by the relative intensity of solar irradiance at that wavelength, generally for air mass 2 (20 degrees solar altitude, at sea level). The products of transmittance and irradiance are summed over the desired wavelength range and then divided by the total irradiance over that range to give the near-normal transmittance. Manufacturers generally specify both the total solar transmittance from 0.3 to 3.0 micrometers and the visible or "daylight" transmittance in the range from 0.4 to 0.75 micrometers. Table 3 shows the normal incidence transmittance values published by two manufacturers.

"Water white" glass containing very low percentages of iron oxide is available from several manufacturers in a limited range of thicknesses and sizes. This has a significantly higher total solar transmittance than ordinary architectural glass but it generally brings a higher price which may offset some of the benefits imparted by the higher transmittance.

The second method of measuring transmittance uses an instrument which employs a sample-holder that can be rotated to vary the angle of incidence between the normal to the glass and the incoming solar rays. With such an instrument the type of data shown in Fig. 2 can be obtained quickly, using a silicon cell sensor for clear glass or a thermopile sensor for glass which has pronounced spectral characteristics. The heavy dots shown in Fig. 3 show points obtained using an algorithm given in Ref. 1, page 26.26, for typical 1/8 in. clear glass. The crosses along the same line represent measurements made at the writer's laboratory. The slight increase in transmittance indicated by the heavy dot at $\theta = 20$ degrees is an anomaly of the algorithm rather than an actual value.

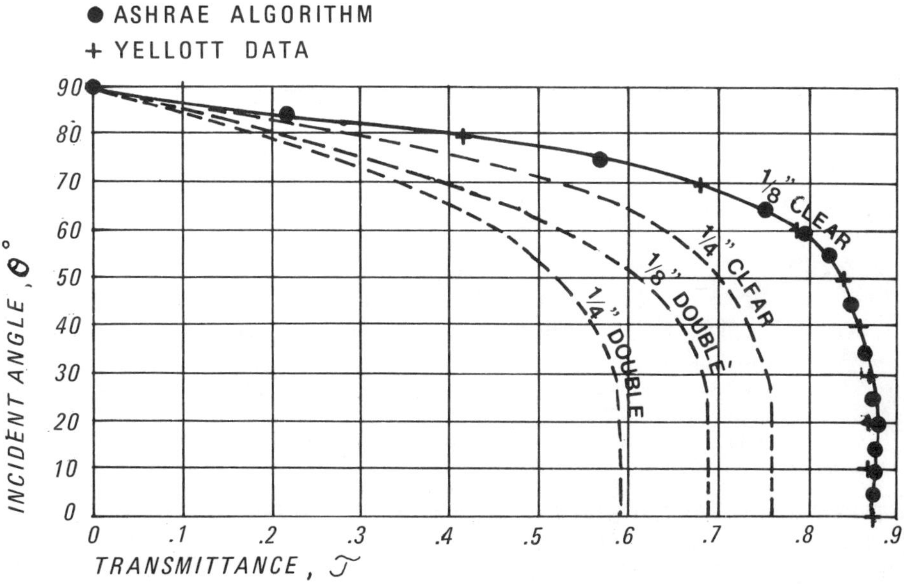

Figure 3. Total Solar Transmittance vs Incident Angle for Single and Double Glazing.

The reason for the rapid drop-off in transmittance for incident angles greater than 30 degrees is the increase in the reflectance which takes place at each air-glass interface. This is about 4% at each interface for values of θ from zero to about 30°, but it increases rapidly as θ rises to higher values above 50°. When θ reaches 90°, the reflectance becomes 100%! This reduction in transmittance is particularly important for south-facing windows in summer when the high solar altitude causes the incident angle on south-facing glazed surfaces to be very high, and their transmittance becomes correspondingly low.

THERMAL PROPERTIES

Although glass is an almost perfect electrical insulator, its thermal conductivity is relatively high, as compared with other common building materials. Values quoted in equally reliable sources differ quite widely, but 7.5 Btu in./hr. ft^2°F appears to be acceptable. The thermal resistance of ordinary window glass thus ranges from 0.017 for 1/8 in. (double strength) to 0.033 for ¼ in. thickness. The units of thermal resistance may be expressed as the temperature difference, in deg F, required to make heat flow at the rate of 1.0 Btu/hr.ft^2 through a 1 in. thickness of a homogeneous material. The thickness unit is sometimes given at 1.0 ft rather than 1.0 in., but in U. S. practice, building material thicknesses are usually specified in inches and so the inch unit is used in this paper.

The specific heat of glass, like most other building materials, is low, and a value of 0.20 Btu/lb.° F is generally used. The weight of glass involved in most building construction is so low that thermal storage in glass is negligible.

A property of glass which is by no means negligible is its longwave emittance. Careful tests made in the U. S., England and Australia have shown that, at the temperatures encountered in buildings, the normal emittance (the emittance for radiation measured along a normal to the surface) for ordinary un-coated glass is close to 0.896[4]. The ratio of total hemispherical to normal emittance for an electrical non-conductor like glass is about 0.94, so a proper value of the emittance for uncoated glass is 0.84. This is nearly 10% lower than the frequently-quoted value 0.93 which was obtained in Germany before World War II using a spectroscope which could not reach the longer wavelengths beyond 15 micrometers. The value given in Ref. 4 was obtained from spectroscopic measurements covering wavelengths from 4.0 to 40 micrometers, which encompasses 98% of the energy emitted by sun-warmed surfaces.

HEAT FLOW THROUGH SINGLE GLAZING

The major resistances to heat flow through single glazing are experienced at the indoor and outdoor surfaces. Because both longwave radiation and convection are involved at each surface, combined coefficients of heat transfer are used to which ASHRAE has long attached values that are nominal rather than precise. For the outer or "weather" surface, the combined coefficient h_o is generally assumed to be 4.0 Btu/hr. ft.^{2}°F in summer, when average wind speed is taken as 7.5 mph; in winter, with a 15 mph wind, h_o is assumed to be 6.0. The resistances are thus 0.25 for summer and 0.17 for winter; at higher wind speeds, R_o can fall well below 0.17 deg F/ (Btu/hr.ft^2).

At the inner surface, two situations may again arise. For most windows, the air at the inner surface responds only to natural convection and longwave radiation; for this situation, ASHRAE uses a nominal inner surface combined coefficient h_i of 1.46 Btu/hr.ft^2.F. R_i is thus 1/1.46 = 0.68.

There is little that can be done about the convection component h_c of this combined coefficient and for vertical windows the following equation is used:

$$h_c = 0.27 \times (\Delta t)^{0.25}$$

where Δt = temperature difference, F, between glass and adjacent air.

This equation was obtained many years ago by combining dimensional analysis (to obtain the exponential relationship) and careful experiments to determine the coefficient, 0.27 for moderately high windows, and the exponent, which is given as 0.25 by some authorities and 0.33 by others.

When forced convection exists because fans or blowers are used, the convection coefficient goes up quite rapidly and more complex relationships are needed (see Ref. 1, p. 26.11). For passive heating and cooling, however, free or natural convection is generally encountered.

Radiation is the major means of transporting thermal energy to or from an inner surface of a window and the basic equation is:

$$q_r = 0.1713 \times E \times [(t_g/100)^4 - (T_w/100)^4] \text{ Btu/hr ft}^2$$

E=effective emittance of glazing and enclosure

T_g, t_w = absolute temperatures of glass and enclosing surfaces, deg R. For two parallel surfaces "infinite in extent", the effective emittance is:

$$E = (e_1 \times e_2)/(e_2 + e_1 - e_1 \times e_2)$$

For a glass surface, e_g = 0.84, and a painted wall, e_w = 0.96,

$$E = 0.806/0.9936 = 0.811$$

The radiation component can also be expressed in terms of a coefficient h_r, which may be found by simply dividing the radiated energy, in Btu/hr.ft^2, by the temperature difference between the two surfaces. Notice that, in radiation calculations, it is the *surface* temperatures which must be used because, in such short distances as those encountered in buildings, air is virtually transparent to radiant energy. In most cases, the indoor wall and ceiling surface temperatures are very close to the adjacent air temperature. By putting the glass temperature equal to 100 ° F, characteristic of a sun-warmed window on a summer day in a moderate climate, and the indoor air temperature at 78°F, the radiation and convection components of heat transfer from glass to air are found to be:

$$h_c=0.27x(100\text{-}78)^{0.25}=0.27x2.3038=0.6220$$
$$h_r=0.1713x0.811x(5.60^4-5.38^4)/(100-78)=0.9199$$
$$h_c+h_r=1.541$$

Thus it is seen that, under these rather typical conditions, the convection component is about 40% of the total while the radiation component makes up the remaining 60%.

This division of responsibility is typical, with radiation taking roughly two-thirds of the load and convection carrying one third. Radiation can be altered to a very substantial extent by changing the emittance of the inner surface of the glazing, and this can be accomplished in several ways. Deposition of a tin oxide coating can reduce the inner surface emittance to as low as 0.20, with a major reduction in h_r and hence h_i. There is generally a significant reduction in total solar transmittance, however, which must, for passive solar heat calculations, be taken into account.

The overall coefficient of heat transmission, U, was adopted many years ago by ASHRAE as the simplest method of estimating steady state heat flow through building components when their characteristics and the indoor and outdoor air temperatures are known:

$$Q=AxUx(t_i-t_o)$$
where A is the area of the building component.

For single glazing, U is readily found since it is simply the reciprocal of the sum of the three resistances:

$$U=1.0/(R_o+R_g+R_i)$$

R_o depends upon the wind speed primarily and so it is 0.25 in summer and 0.17 in winter. R_i depends upon the glass and air temperatures and the glass emittance, as shown above, but for uncoated glass it is usually close to 0.68. For ¼ in. glass, the resistance is very low and it can often be neglected, so, for a first approximation, U = 1.0 / (0.25 + 0.68) = 1.08 for summer. If the resistance of ¼ in. glass, 0.03, is added in, U = 1.04 Btu/hr.ft^{2}F, which agrees with the ASHRAE value given in Fig. 4.

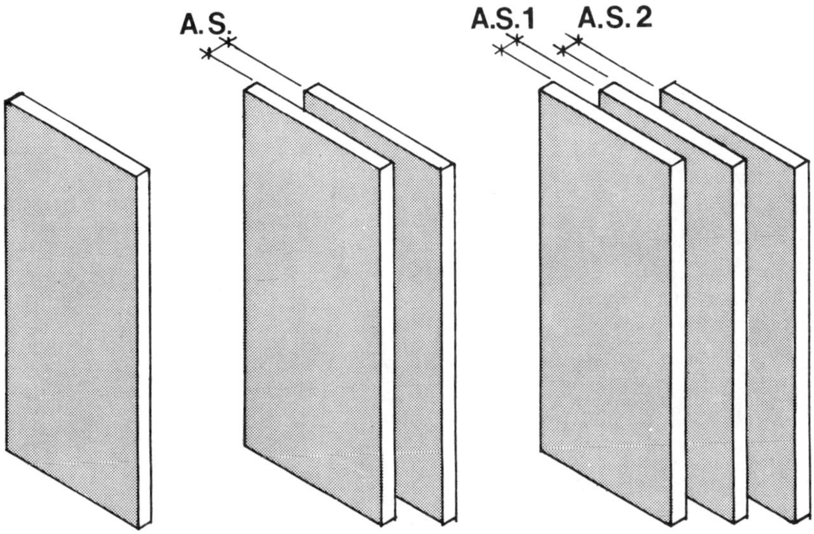

U–VALUES

Type	Single	Double		Triple	
AIR SPACE		1/4″	1/2″	1/4″	1/2″
Summer	1.04	0.61	0.56	0.44	0.39
Winter	1.10	0.58	0.51	0.39	0.31

Heat transfer rates (BTU/hr. ft^2 F) for single and multiple glazing as a function of season and air space size. U values are for glazing only and do not include frame or sash effects.

Figure 4. Architectural Glazing Types and Their U-Values.

In winter, U = 1.0 / (0.17 + 0.03 + 0.68) = 1.14 which is slightly higher than the ASHRAE value shown in Fig. 4. That value is based on an inner surface coefficient of 0.709, which is about 4% higher than the conventional 0.68.

HEAT FLOW THROUGH MULTIPLE GLAZING

For double and triple glazing, the air spaces between the glasses add significantly to the thermal resistance and so the U-values are substantially lower than for single glazing. As the 1978 paper showed, the reduced winter heat loss is somewhat offset by the reduction in heat gain due to the lowered solar transmittance. Consequently, a generalized statement that double or triple glazing is always justified might well be incorrect.

CONCLUSIONS

Glass is an unusual building material in that it can both admit and dissipate heat, depending upon the direction of the indoor-outdoor temperature differential, and it can admit solar heat when it is needed. Glass, however, has its own unique set of properties and these must be taken into account when passive solar buildings are being designed. Fortunately, procedures are now available by which both solar and temperature-difference heat flows can be estimated with a good degree of accuracy and the ASHRAE Handbook of Fundamentals (Ref. 1) gives these in great detail. The present paper gives a brief look at some of the non-solar design issues which are too often overlooked.

REFERENCES

1. *1977 Handbook of Fundamentals*, American Society of Heating, Refrigeration and Air-Conditioning Engineers, New York.

2. *Architectural Data Handbook*, Fifth Edition, PPG Industries, Pittsburgh, PA. 1965.

3. Baumeister and Marks, *Standard Handbook for Mechanical Engineers*, Seventh Edition, EcGraw-Hill, Inc., New York, 1967, p. 4.9.

4. Yellott, J. I., "Selctive Reflectance — a New Approach to Sun Control," *ASHRAE ·Transactions*, Vol. 69, 1963, pp. 418-438.

Air Leakage Through New Residential Windows

John L. Weidt and Jenny Weidt
John Weidt Associates
Chaska, Minnesota

INTRODUCTION

Recent events underlining the dependence of the United States on dwindling fossil fuel energy resources have increased the need to identify areas in which wasteful energy use can be curtailed. The use of passive solar design strategies is becoming popular as a means of diminishing our requirements for the use of fossil fuel resources for heating and cooling buildings. Builders, designers and homeowners are becoming increasingly interested in structures that work with the environment and effectively utilize renewable energy sources.

An important part of developing an energy-efficient building system is understanding the performance capabilities of the building's components, thereby assuring that they will perform harmoniously as a system rather than counteracting each other. Windows are building components which need a great deal of planning to obtain peak performance within such a system.

Historically, the energy effectiveness of windows has been given little consideration in the design and construction of buildings; as the concern for effective energy utilization has expanded, the importance of energy efficient windows has become more evident. Windows have positive attributes such as the ability to collect desired heat to warm a building in winter, to provide light in the building year-round and in the case of operable windows, to allow for natural ventilation and thus natural cooling of the interior environment. These same positive attributes of windows, if misutilized, can lead to excessive energy consumption, such as unwanted heat gain during the summer months which increases cooling loads and uncontrolled air infiltration—air leakage—which robs the interior space of warm air during the winter months. This focuses on air leakage, a *potential* negative attribute of operable windows, and hopefully will provide some useful insights for designers, specifiers, builders and homeowners. The potential negative impacts of air leakage through windows can be minimized through the proper selection and installation of the window unit, thus allowing the window's useful energy control functions to be maximized.

This paper discusses the critical air leakage factors which must be considered when designing, specifying or installing new windows. It outlines results obtained when the air leakage of a number of newly installed windows was measured in the field and compared to laboratory test data. It also describes the variables which were found to have the greatest impact upon window air leakage, and what to look for to select a window with maximum performance. Finally, it describes the results of some additional tests made on fixed windows, on window installation, on tests run on windows in cold weather, and on windows which had been installed some years earlier.

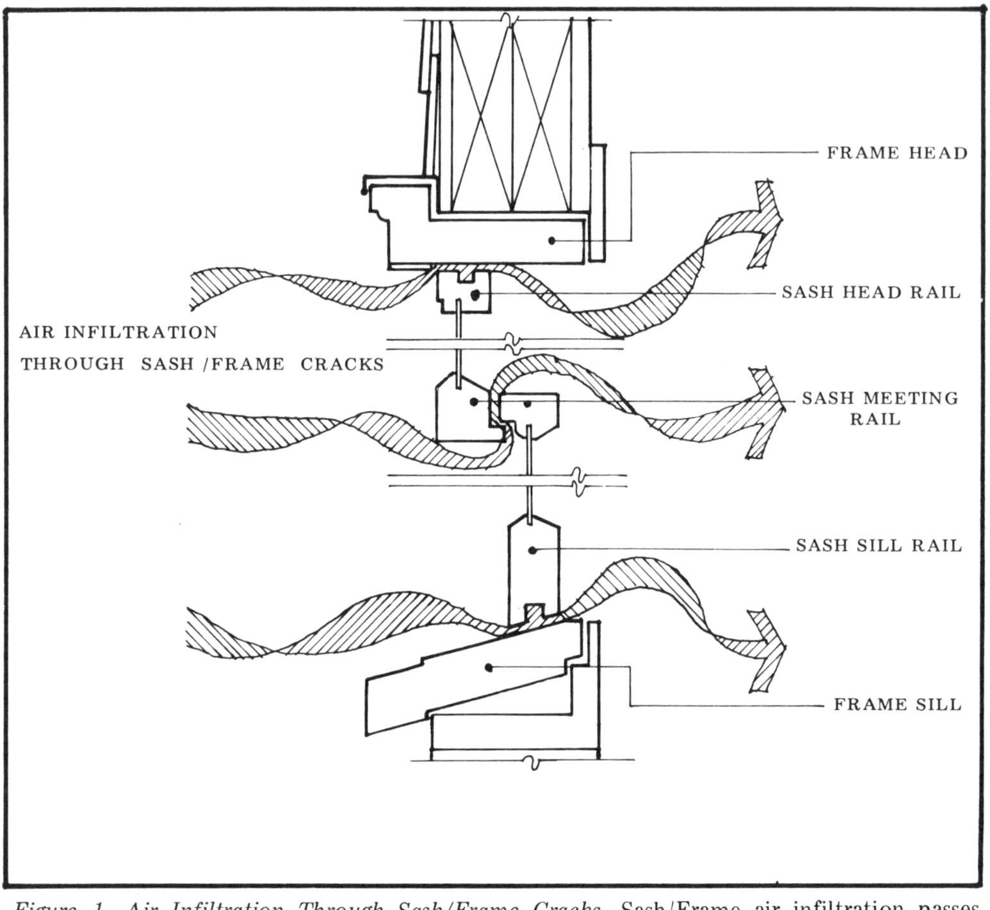

Figure 1. Air Infiltration Through Sash/Frame Cracks. Sash/Frame air infiltration passes through the cracks between various window sash parts and between the sashes and frame.

Air infiltration occurs when pressure and temperature differences between the exterior and interior of a building induce air leakage through its envelope (Figure 1). Prime locations for this leakage are the cracks between the various parts of the window unit such as between the sash and frame. Determining the amount of air passing through these locations in the window unit was the subject of a recently completed study by the Minnesota Energy Agency for Lawrence Berkeley Laboratory and the U.S. Department of Energy. The firm, John Weidt Associates, was the prime sub-contractor. The project tested the air leakage performance of windows in the field and investigated the reasons for these performances. The testing was performed in Minneapolis/St. Paul, Minnesota in 1978 and 1979 and focused on determining the air leakage performance of new residential windows installed in new construction. The major operation types tested were casement, double slider, double hung, single slider and single hung windows (Figure 2).

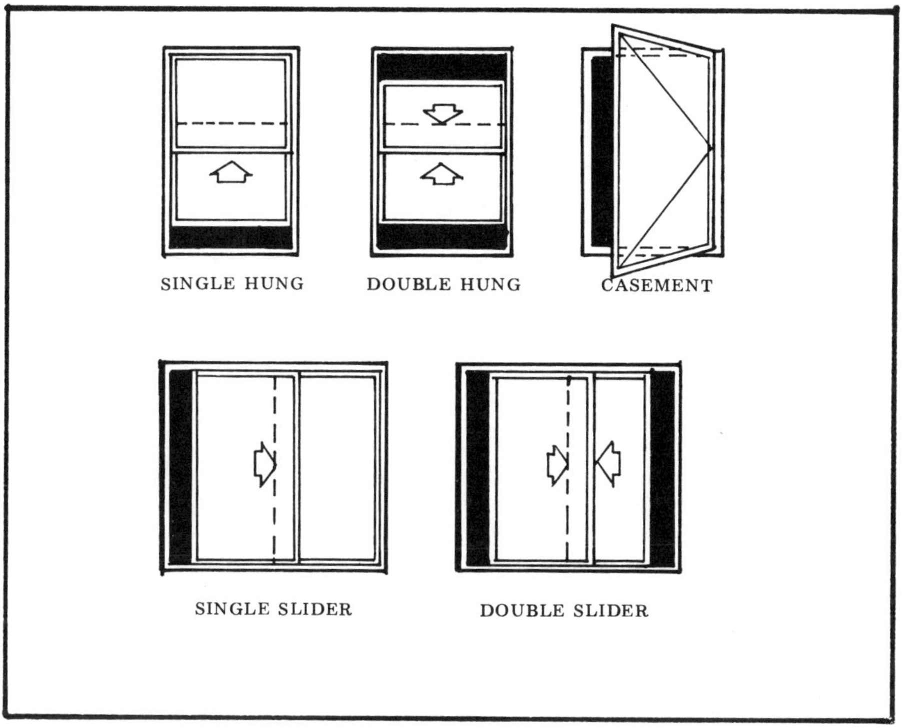

Figure 2. Window Operation Types

The data base was selected to test a typical cross-section of windows being installed in the Twin Cities area. In all, the air leakage performance of 192 aluminum, wood or clad wood windows manufactured by 16 different companies was measured. All major operation types produced by each manufacturer were tested. The air leakage was measured with a portable testing device which allowed windows to be individually pressurized and resultant air leakage measured. In addition to measuring the air flow, 32 other variables were recorded for later analysis. The data obtained in the field were then compared to laboratory measurements of air leakage performance.

It is important to recognize that no regulations or standards exist that mandate specific performance requirements of a window once it has been installed. Window associations such as National Woodwork Manufacturers Association (NWMA) and Architectural Aluminum Manufacturers Association (AAMA) have certification requirements that a window, when tested in a laboratory, perform within the specified maximum limit of .50 cubic feet of air per minute per lineal foot of crack (cfm/lfc). A number of public standards such as National Council of States on Building Codes and Standards (NCSBCS) Model Energy Code, American Society of Heating Refrigerating and Air-Conditioning Engineers (ASHRAE) Standard 90-75, Housing and Urban Development Minimum Property Standards (HUD MPS) and Federal Housing Administration Minimum Property Standards (FHA MPS) require certification of a product line through laboratory testing. Manufacturers frequently refer to these air leakage test results in their advertising by either stating that they meet or exceed the .50 cfm/lfc standard or by publishing the actual laboratory test results for a particular model of their product line which may report air leakage rates far less than this .50 cfm/lfc; sometimes as low as 0.01 cfm/lfc.

Unfortunately, there is a general misconception about how these standards and test results can be used. Designers and builders who specify and purchase windows for construction frequently rely on the laboratory test results to be indicative of the window's field performance capabilities. Little actual test data are available that relate the laboratory performance of a window to its installed performance. The results of this study clearly indicate that the field performance of a unit can differ significantly from its laboratory performance.

RESULTS

The industry and government standards of maximum air leakage of .50 cubic feet of air per minute per linear foot of sash crack were used as the primary reference for comparison of the field results. A second set of references were manufacturers' reports of air leakage performance obtained in the laboratory. The distribution of the results obtained by field testing when compared to these references were interesting. The windows tested had a wide range of performance; from less than .01 cfm/lfc to greater than 2 cfm/lfc (Figure 3, Table 1). When comparing this range of performance to the references, it was found that 40% of all the windows tested in the field did not meet the laboratory test requirement of .50 cfm/lfc and that 60% of the windows tested in the field did not meet the manufacturers' reported laboratory performance (Figure 4). The average air leakage performance of casement windows was .23 cfm/lfc, double sliders was .61 cfm/lfc, double hung was .72 cfm/lfc, single sliders was .79 cfm/lfc, and single hung was .96 cfm/lfc (Figure 5, Table 2).

Casement windows far out-performed sliders, and sliders generally out-performed hung windows, irrespective of all other observed variables, such as the sash and frame material, the manufacturer of the window, or the installer of the window. Manufacturers who made casement, slider and

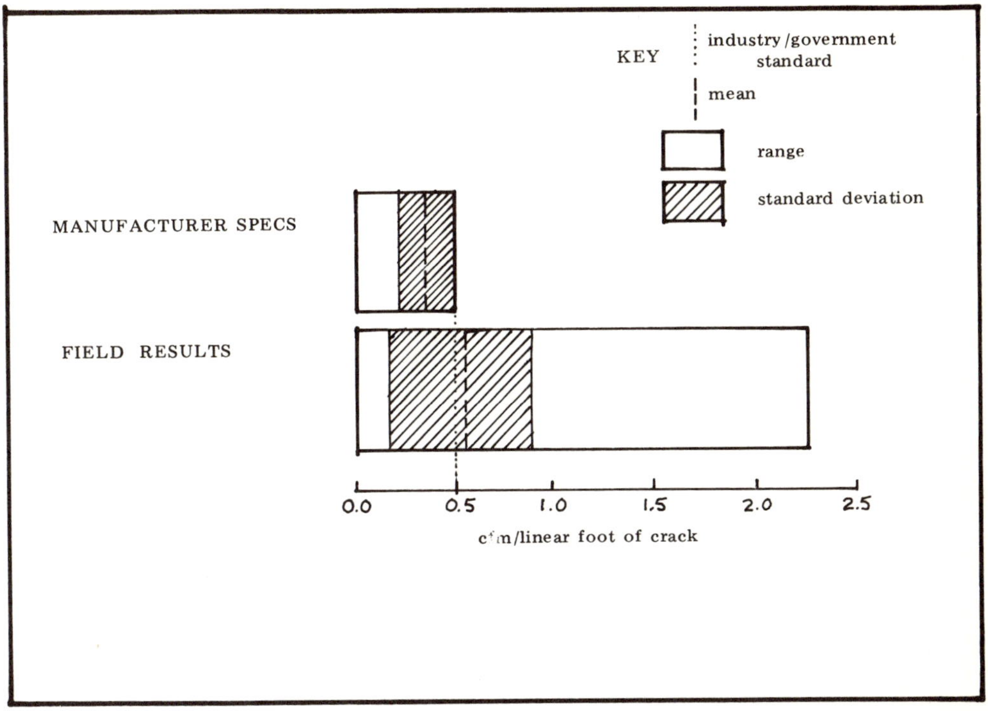

Figure 3. Air Infiltration of Installed Windows. The graphs compare the range and mean of manufacturer's published air leakage rates and industry/government standards for air leakage with results of tests performed on installed windows.

TABLE 1
Air Infiltration of Installed Windows

Description	Range	Standard Deviation	Mean
Industry/Government standards: .50 cfm/lfc			
Manufacturer Specifications	.04 - .50		
Field Results	.008 - 2.279	.130 - .920	.524

hung windows generally produced casement windows with lower air leakage rates than their sliders, while their slider windows generally had a lower air leakage rate than their hung windows. In general, any of the manufacturers tested could make a casement window that out-performed their own and other manufacturer's slider or hung windows, and even a sloppy installation of a casement window appeared to out-perform a very well installed slider or double hung window.

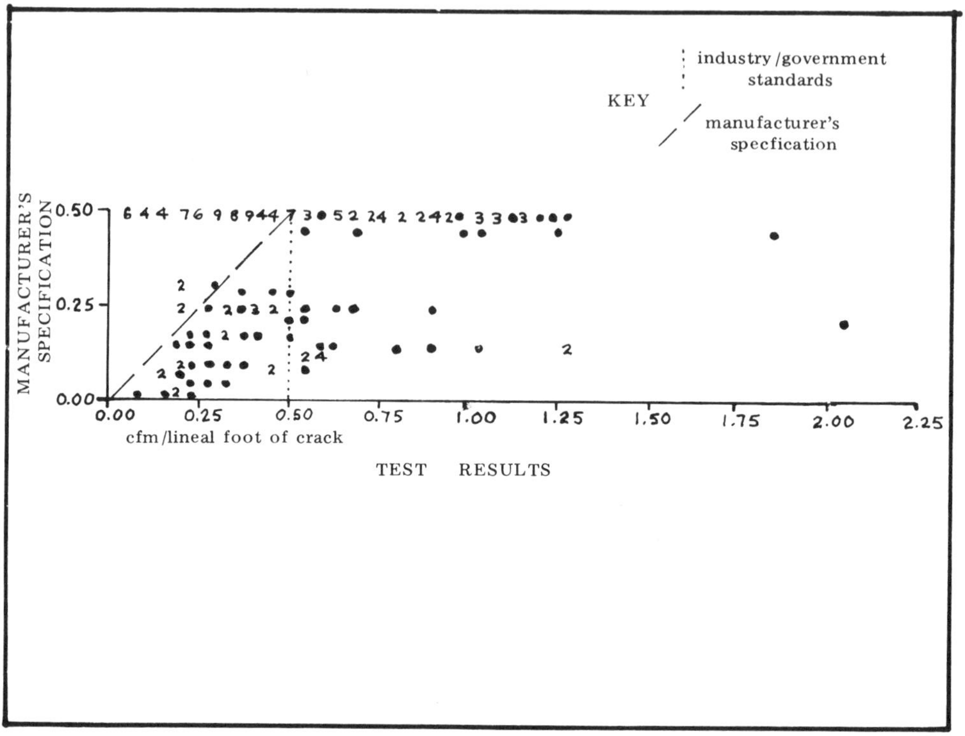

Figure 4. Scattergram of Field Results. Results of each test are plotted relating to their actual performance to manufacturer's specifications. A number indicates that more than one result occurred at a given point. As an example, a point occurring at X=.75, Y=.25 means that the window manufacturer reported a lab test at .25, the field-measured leakage of the unit was .75. Points to the left of the reference lines relate to windows whose field air leakage was less than reference, those to the right had air leakage rates greater than reference.

The air leakage of the windows tested followed certain patterns which were at times related to physical defects in the unit. These air leakage patterns varied from operator type to operator type, but were somewhat similar within each operator type. For instance, casement windows generally leaked at their corners, sliding windows leaked at the corners and at the meeting rails, and double hung windows leaked at all of the above points and at the sills (Table 3). The reasons for these particular areas of leakage appeared to be due to one or more of the following physical flaws in the window:

1. A break in the weatherstrip continuity around the window was a common problem— either the weatherstrip was cut at a corner, misapplied at a corner, or changed planes in a corner. In several cases the window was designed so that the vertical and horizontal pieces of weatherstripping were installed in planes that could not meet and thus there was a gap of the weatherstrip at that juncture.

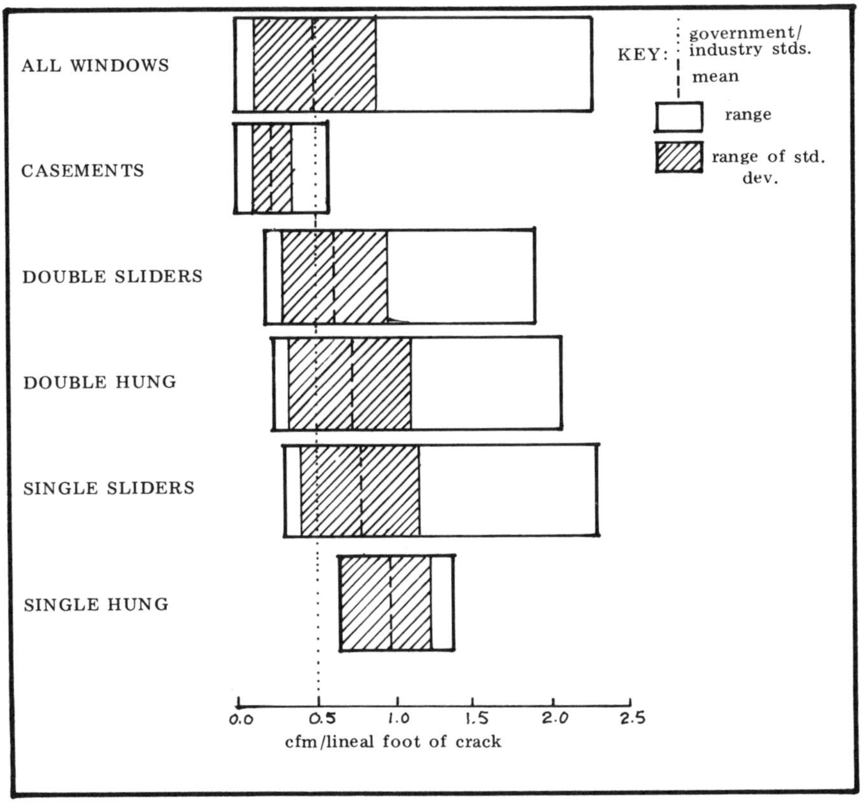

Figure 5. Comparison of Window Types, Field Results.

TABLE 2
Field Results: Comparison of Window Types

Window Type	No. of Tests	Range	Range of St. Dev.	Mean
All Windows	192	.01–2.28	.13–.92	.52
Casements	79	.01–.58	.11–.35	.23
Double Sliders	33	.17–1.90	.27–.96	.61
Double Hung	38	.22–2.06	.31–1.14	.72
Single Sliders	31	.30–2.28	.38–1.19	.78
Single Hung	11	.68–1.37	.67–1.25	.96

TABLE 3
Performance of Windows with Defects

Window type	Number of windows with one or more known defects	Air leakage performance of windows with defects to average performance within window type
Casement	28%	45% Above Average
Double Slider	45%	Average
Double Hung	82%	7% Above Average
Single Slider	65%	12% Above Average
Single Hung	55%	24% Above Average

2. There were problems with sash fit, particularly in sliding and hung windows. In a few windows, out-of-square sash allowed excess leakage in the diagonal corners of the operable portion of the window. Sash with loose fit did not provide good, tight fitting weatherstripping and thus had excessive leakage along the length of the sill or meeting rail.

3. Locking hardware was also a problem with some window types where the locking devices, rather than drawing the sash together, would push the sash apart.

Some window units which had been tested during the new window portion of the program were retested during the winter months to investigate the possibility of poorer performance during cold weather. The average outdoor temperature encountered during the initial testing was 60.2 F, while the average outdoor temperature encountered during cold weather retesting was 7.8 F. The results obtained indicate no significant change in the air leakage performance during cold weather. After each opening and closing of a window unit, its performance will vary since factors such as dirt and the care of the operator in closing the unit can greatly effect its seal; the difference of the two results (original versus cold weather) could be attributed to the variation in performance that is to be expected with normal operation of the unit. (Figure 6, Table 4).

The ability of a window to maintain its air infiltration performance over a period of years is an important energy conservation attribute. In this project, ten tests were made on windows which had been installed for from 2 to 8 years, and for which manufacturers' specifications from the installation year were available. The performance of 7 of the 10 older windows tested compared favorably with the performance of new windows of the same operation type, although none performed as well as may have been anticipated from the manufacturer's specifications (Figure 7, Table 5).

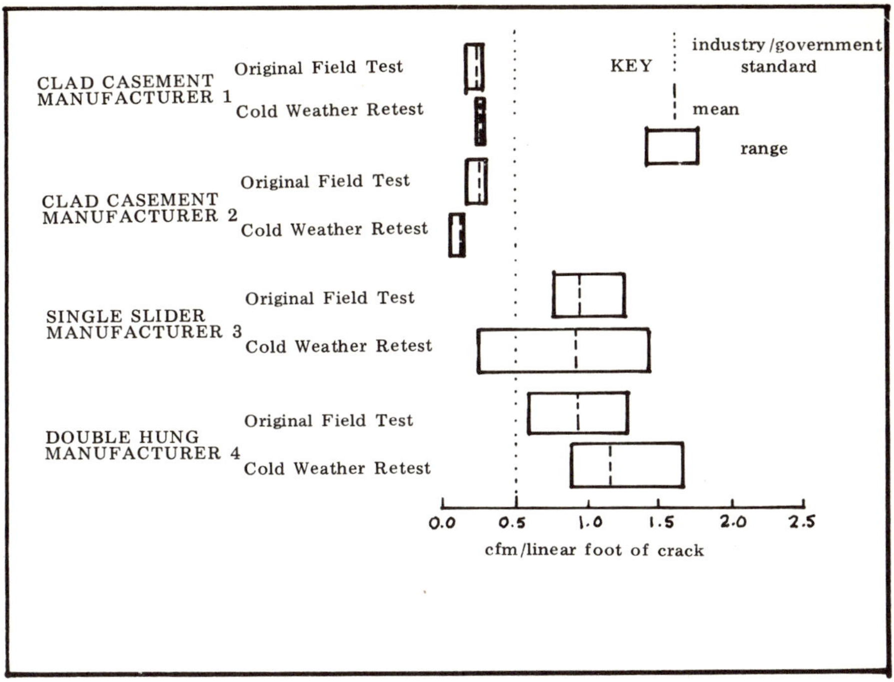

Figure 6. Cold Weather Retests. A sampling of windows which had been tested during the summer months were retested in the winter to observe performance differences.

TABLE 4
Cold Weather Retests

Manufacturer	Window Type	Original/Retest	Range	Mean
1	Clad Casement	Original	.16–.27	.20
		Retest	.24–.27	.25
2	Clad Casement	Original	.01–.49	.14
		Retest	.14–.17	.16
3	Single Slider	Original	.62–2.28	.99
		Retest	.54–1.41	.91

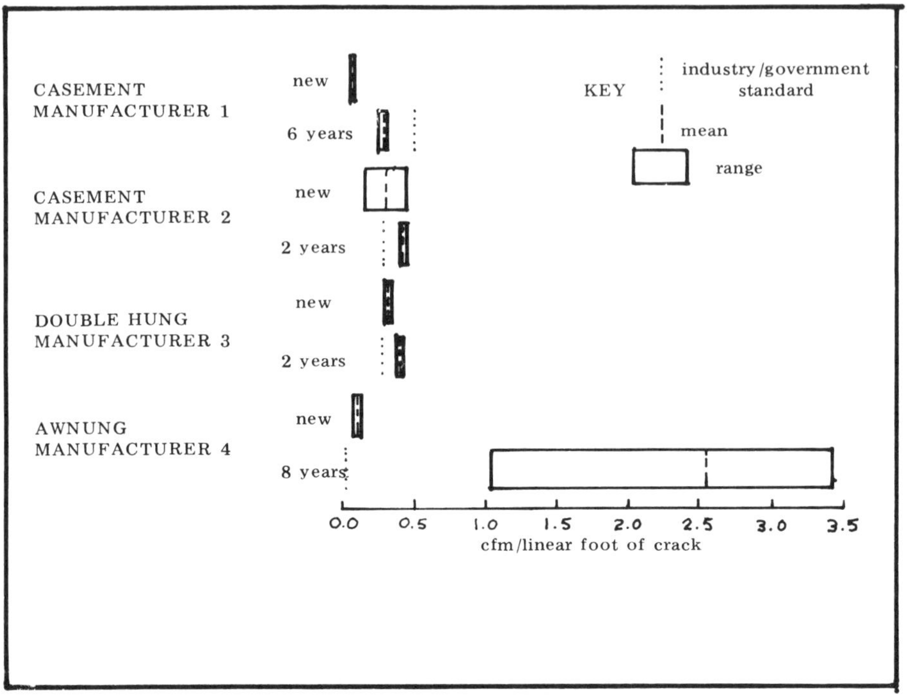

Figure 7. Performance Over Time. A relation of the performances of new windows to their counterparts two to eight years older.

TABLE 5
Performance Over Time

Manufacturer	Window Type	New/Over Time	Mfgr. Spec.	Range	Mean
1	Casement	New		.04–.08	.05
		Over Time	.50	.23–.32	.27
2	Clad Casement	New		.15–.46	.31
		Over Time	.25	.39–.46	.43
3	Double Hung	New		.31–.39	.40
		Over Time	.25	.41–.44	.43
4	Awning	New		.10–.15	.12
		Over Time	.04	1.17–3.42	2.56

There was substantial degradation of performance in 3 of the older windows tested, which were awning units. The sash was equipped with a roto operator at the center of the right jamb and a locking mechanism at the center of the left jamb. Ice, a result of condensation, built up between the sill stop and the sash and as it expanded, pushed the sash outward until the teeth on the roto operator snapped and the sash warped. Although only 3 of these units were tested, this condition was observed in a total of 22 similar window units.

Six fixed window units were tested for their installed air leakage performance. The performance of the units tested ranged from .12 to 2.04 cfm per square foot of sash area (cfm/sf), exhibiting relatively poor air leakage characteristics (figure 8, Table 6). Leakage of the units was located with smoke and usually occurred near corners both at the glazing/sash and the sash/frame interface. The poorest performer did not appear to have continuous sealant between the glazing and the sash as a strip of cardboard could easily be inserted between the sash and glazing at several locations.

Currently, air leakage performance is expressed in terms of cubic feet per minute of air flow per linear foot of crack. Normally the selection of a window is not made on the basis of its linear feet of crack and so the designer or buyer needs to do some translating to get an understanding of the total volume of leakage to be expected. Evaluation of the window's air leakage performance should not be made solely on the basis of leakage per linear foot of crack, because the total volume of air leakage for certain types of windows which have inherently longer crack length (double slider and double hung) is considerably higher than for other windows. Crack length can be misleading if used as criteria when selecting between two window types such as single sliders and double sliders. These two types of windows may have exactly the same overall dimensions and capacities for ventilation. When the air leakage is calculated via the crack length method, and the leakage per linear foot of crack is the same, the double slider will appear to equal the performance of the single slider when, in fact, the total volume of air infiltrating through the double slider will be over 60% greater due to its additional crack length.

Technical differences in the definitions of single and double operating units can lead to additional confusion. For instance, some side sliding windows observed during the field testing appeared to be double slider windows as both sashes were equipped with hardware and track; one sash, however, was held in place with set screws. The manufacturer defined this window as a single slider. Other side sliding windows were observed in which only one sash was equipped with a full width track, the track for the second sash extended to only one half the width of the frame and the sash had no hardware (handles, etc.) for operation. Nonetheless, the sash was unrestrained—its manufacturer defined the unit as a double slider. Identification and performance calculations of all windows tested in this project were based upon the manufacturer's definitions of window types;

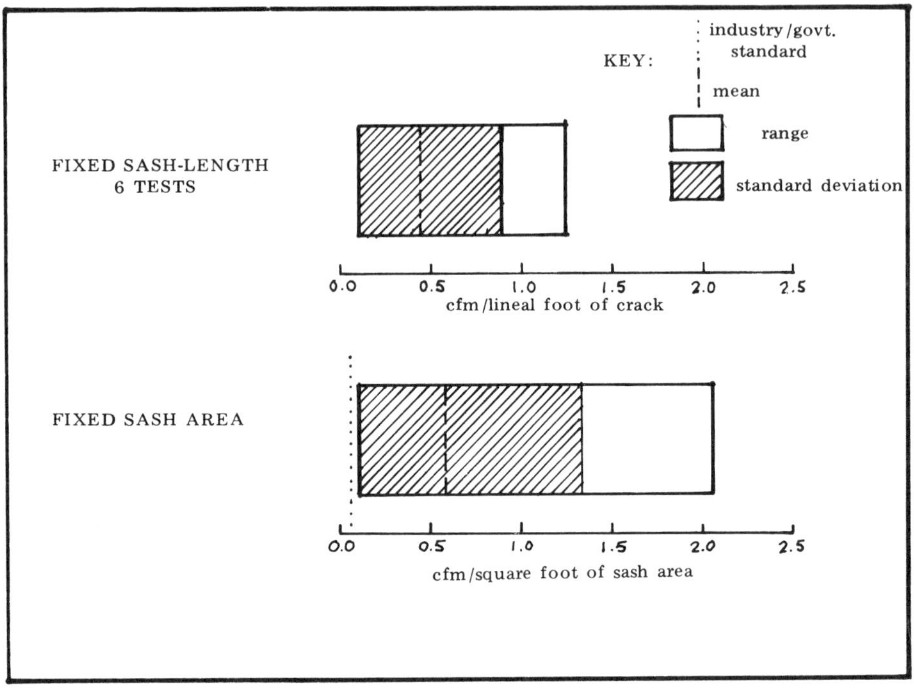

Figure 8. Results of Testing Fixed Sash Related via Crack Length and Sash Area.

TABLE 6
Results of Fixed Sash Related via Crack Length and Sash Area

Method of Calculation	Range	Range of Std. Dev.	Mean
cfm/lfc	.11 – 1.21	.11 – .81	.39
cfm/sf	.12 – 2.04	.12 – 1.34	.60

thus the air leakage of these two windows was related to the appropriate crack length for single and double slider windows, respectively. In analyzing the performance of these two windows, the "double" slider out-performed the "single" slider on the basis of air leakage per linear foot of crack (the double slider had 69% more crack length and thus the total air infiltrating through the unit could be divided by this substantially greater crack length). When these two windows of equal area were compared on a total leakage basis, however, the single slider substantially out-performed the double slider (figure 9).

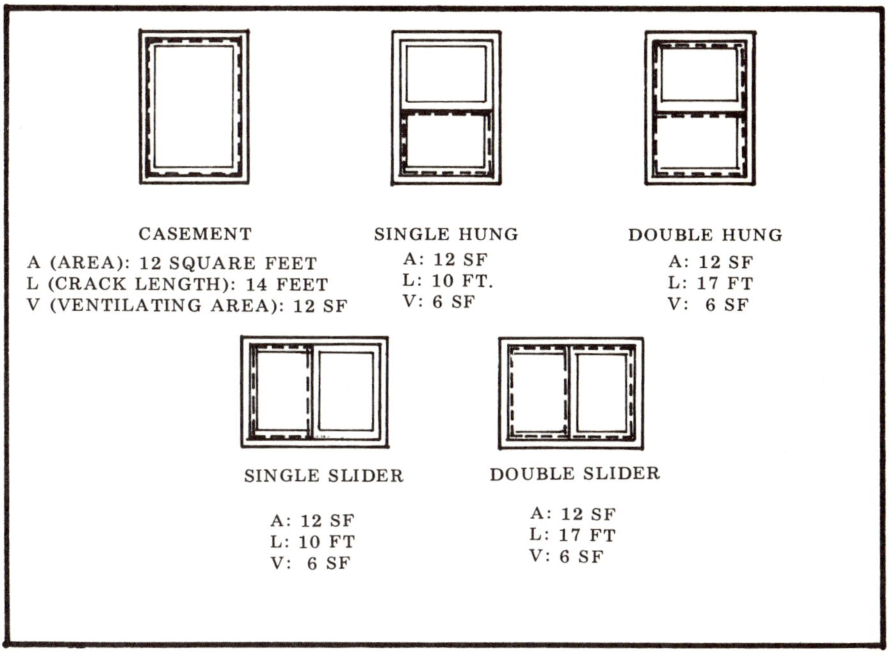

Figure 9. Comparison of Crack Length, Sash Area, and Ventilating Area of Typical Operation Types.

The performance of the major operation types measured in this study were expressed in terms of air leakage per linear foot of crack, per square foot of sash area, and per square foot of free ventilating area (Figure 10, Table 7). Large shifts in relative performance, dependent on the expression of leakage used, can be observed, particularly in the following areas:

SINGLE SLIDER RELATIVE TO DOUBLE SLIDER WINDOWS:
When air leakage is expressed per linear foot, the leakage of the double slider window is 72% as great as the single slider. When leakage is expressed as a function of either sash or vent area, the performance of the single slider is 62% and 60% that of the double slider, respectively.

SINGLE HUNG RELATIVE TO DOUBLE HUNG WINDOWS:
When air leakage is expressed per linear foot, the leakage of the double hung window is 66% as great as the single hung window. When leakage is expressed as a function of either vent or sash area, the performance of the single hung window is 81% and 84% that of the double hung window, respectively.

CASEMENT WINDOWS RELATIVE TO ALL OTHER WINDOWS:
Whether the leakage is expressed per linear foot, per unit sash area or per unit ventilating area, the average casement window out-performs the average of the next best performing window operation type.

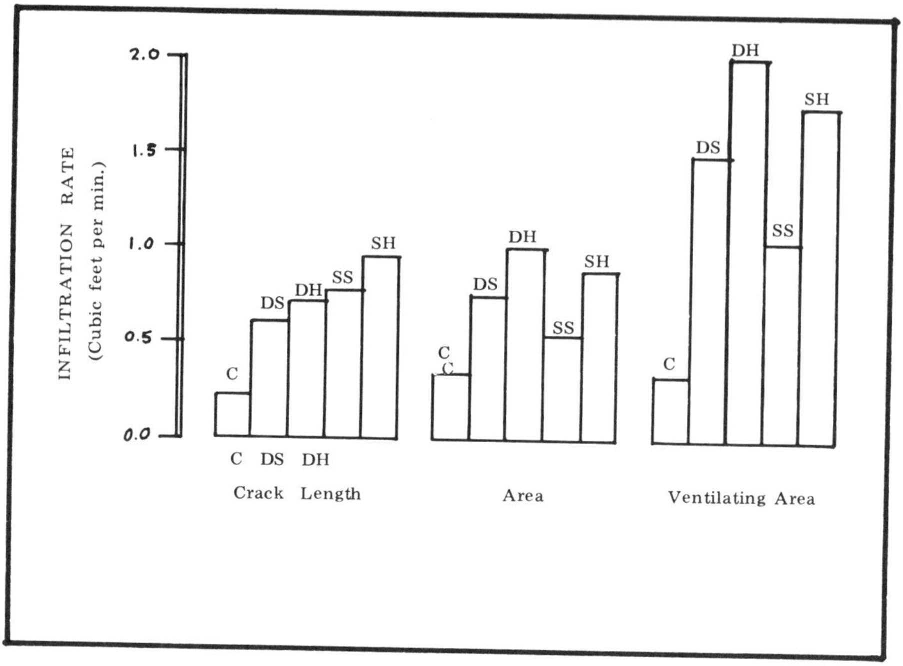

Figure 10. Expression of Air Leakage Rate. The results of testing new windows in new construction were calculated via the three methods displayed above. *Figure 11. Air*

TABLE 7

Expression of Air Leakage Rate

Method of Calculation	Casement	MEAN RESULTS Double Slider	Double Hung	Single Slider	Single Hung
cfm/lfc	.23	.61	.72	.78	.96
cfm/sf	.34	.76	1.015	.55	.88
cfm/vsf	.34	1.57	2.097	1.14	1.77

An important consideration in evaluating the air leakage performance of the total installed window unit is the amount of air passing between the window unit and the wall opening (Figure 11). The project undertook to obtain a better understanding of the magnitude of this leakage through application of some experimental testing techniques. Although the techniques proved imperfect, they nonetheless can be used in a qualitative way to determine the importance of proper installation of the window unit in a wall on energy performance.

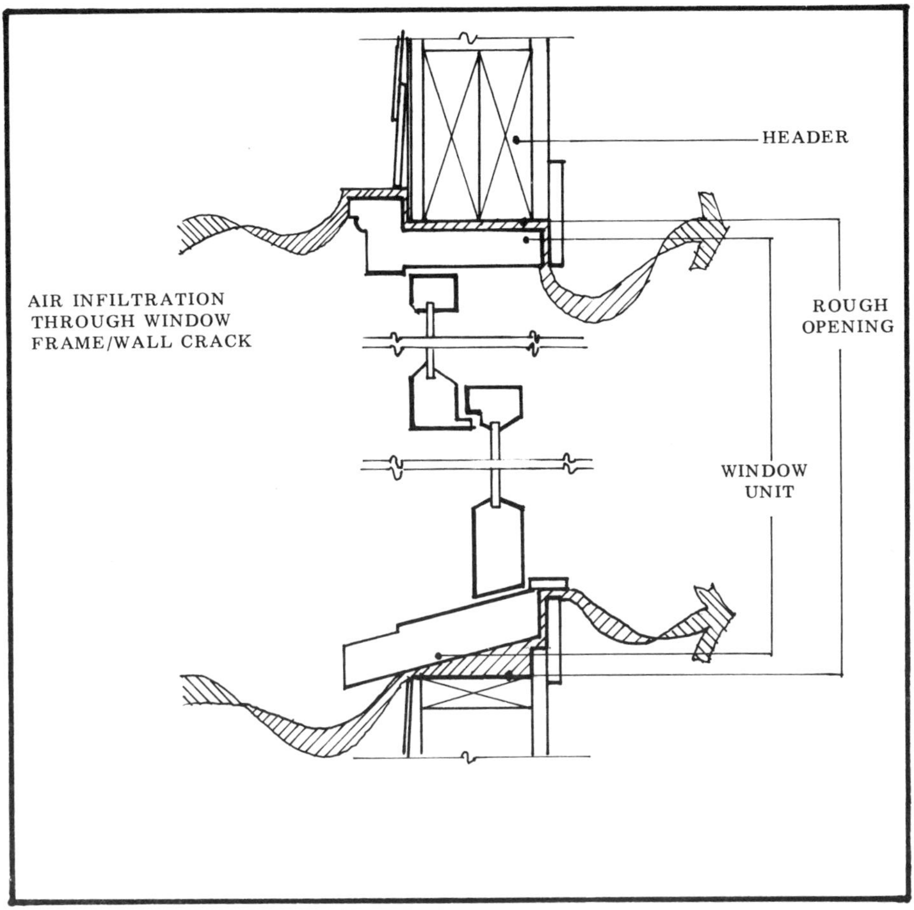

Figure 11. Air Leakage Through Frame/Wall Cracks. Frame/wall air leakage occurs at the cracks between the window frame and the wall.

It is interesting to note the relative impact of total air leakage as measured through the frame/wall crack when compared to the leakage through the sash/frame crack of the installed window unit (Table 8). Although the test method allows some degree of imprecision in the absolute measurement of the volume of outside air flowing through the sash/frame crack, it nonetheless indicates that the air leakage performance of the crack between the window unit and the wall has a significant effect on the air leakage performance of the entire window unit as installed. The air flow measurements at the frame/wall crack indicate that air leakage at this location can be nearly that of the window unit itself. Control of air leakage at the frame/wall crack is relatively easy during window installation but cannot be controlled easily after the window is installed. It is, therefore, imperative that window installers pay close attention to limiting air leakage at the frame/wall crack. The frame/wall crack leakage, although significant, was not found to have the impact of the sash/frame leakage of the window unit itself in this limited series of tests.

TABLE 8
Frame/Wall Leakage Area

Test No.	Air Leakage at the Frame Wall Crack		Air Leakage at the Sash/Frame Crack	
	Per Foot	Total	Per Foot	Total
1/1	.43	10.71	.25	10.9*
1/2	.24	5.91	not measured	
1/3	.38	6.61	.51†	10.2†
1/4	.23	4.09	.45†	8.9†

* Total leakage through combination of operator and fixed sash
† Figures from similar units tested as installed by same contractor

SUMMARY

In summary, it is important to remember that windows do not necessarily perform per laboratory specification when installed in a building and thus their selection should be based on a number of other factors:

1) The most important single variable effecting a window's performance is its operation type. A casement window significantly outperforms a slider window or hung window on the average.

2) Look for manufacturing flaws in the windows under consideration. Check to see that hardware operates properly and that weatherstrip is complete and properly installed.

3) Select windows on the basis of total volume of air leakage related to the design use of the window. Keep in mind that windows with a high ratio of crack length to glazed or ventilating area will have correspondingly high air leakage performance. Ask for additional documentation for claims of very low air leakage performance.

4) Take care in the installation of the window. Careful sealing of the window unit into the rough opening will insure that minimum air leakage will be allowed between the window unit and the wall.

This field study has demonstrated that the negative air leakage aspects of windows can be controlled with a better understanding of how and why air leakage takes place. Through the proper design, selection and installation of windows, this air leakage can be minimized.

The work described in this report was funded by the Office of Buildings and Community Systems, Assistant Secretary for Conservation and Solar Applications of the U. S. Department of Energy under Contract No. W-7405-ENG-48 with the Lawrence Berkeley Laboratory. This paper is a summary of the original project report, *Air Leakage of Newly Installed Residential Windows*, LBL Report No. 11111, available from Energy Efficient Windows Program, Bldg. 90-3111, Lawrence Berkeley Laboratory, Berkeley, CA 94720.

Slice of the Sun

Marietta S. Millet
Department of Architecture
Gould Hall JO–20
University of Washington
Seattle, Washington 98195

A great American poet once asked the architect, "What slice of the sun does your building have? What light enters your room?" - as if to say the sun never knew how great it is until it struck the side of a building. ...Louis I. Kahn[1]

Just as the design of buildings which responded to climate was diverted by the availability of inexpensive and "reliable" fuels in the last thirty years, likewise designing for daylight was eclipsed by the easy availability of electric power. Suddenly the designer had the power to predetermine and set the interior luminous environment without having to understand or account for the daily and seasonal changes of daylight. While daylight was always considered an imperative element in residences, along with view, natural ventilation, and connections to the outside, "windowless schools" appeared along with other building types which minimized contact with external influences in order to artificially control the internal environment. And yet, it is in these medium and large-sized buildings peopled during daylight hours that the utilization of daylight can save substantial amounts of electric energy and operating costs, provide excellent lighting quality, and reduce peak loads on electric utilities. Many issues must be recognized, understood, and resolved in order to realize fully the potentials of daylighting working in concert with energy-conscious design. Some of these issues are presented along with recent developments and design approaches.

INTRODUCTION

That part of building design which we call "daylighting" is far more than a technical issue: it is the essence of design, that which determines how a building will be perceived and experiences visually. Our most revered architects, both historically and in the recent past, are known as skillful manipulators of light and lighting effects, from Borromini through Wright, Aalto, and Kahn.

They have also written about it explicitly, Le Corbusier: *Architecture is the masterly, correct, and magnificent play of masses brought together in light.*[2]

And Kahn:

No space, architecturally, is a space unless it has natural light. Natural light has varied mood of the time of the day and the season of the year. A room in architecture, a space in architecture, needs that life-giving light—light from which we were made.

We turn to the work of these masters in order to learn how to apply basic principles of good architecture in practice. But it is difficult to interpret historical forms literally because the design process is now mediated by new technologies. However, there are still many old lessons to be incorporated into the new vocabularly of energy-conscious design we are trying to forge, and some of the most striking images are of light. Passive solar building design is not new ("there is nothing new under the sun"). We may use an updated and upbeat terminology, we may have new products, we may even have radically different design approaches, but somewhere, sometime, someone has tried out most of the approaches we are now trying—maybe only partly, maybe for different reasons, maybe in different applications—but we stand to learn much from what has gone before. [3]

The term "daylighting" as used in this paper encompasses what I consider to be the whole range of daylighting design. Daylighting cannot be tacked on as a separate engineering package but must be included in the beginning design concept. Daylighting design includes the design of the supplementary electric lighting system and, as an extension, the total electric lighting installation. The lighting fixture layout must be designed both to complement the daylighting and also to operate as an independent system after dark. In order for this to be accomplished, the lighting control system must be an integral part of the design. The total lighting system is "dynamic," with daylight and electric lighting working separately and together at different times in response to changing time and weather and the varying activities of the inhabitants. A good control system allows adjustment to accommodate these changes. We cannot separate daylight and electric light—but we can manipulate them artfully to work well together.

In the first section of this paper, the influence of climate on daylight design will be discussed briefly. Next, some daylighting issues related to energy conservation in large buildings will be presented. A brief review of available design aids is presented in the third part, and in the final section some new directions in daylighting design are discussed.

INFLUENCE OF CLIMATE

Daylight is a dynamic function of site, changing with the latitude, dominant weather patterns, the seasons, local atmosphere, the diurnal cycle, and specific site conditions. The relationship between daylight and useful interior illumination is affected by all the factors listed above, which can loosely be called climate. Since the climatic variations in the United States encompass most if not all of the distinct climate zones of the world, we have an expansive range of indigenous architecture to study to understand the interplay of light and climate characteristics. While it is possible to make generalizations about daylighting approaches in indigenous architecture in response to climate, there is never a hard and fast rule. Many other factors enter in: available materials, social and organizational structure of the building occupants, use of the building, and cultural patterns, to name a few. Nevertheless, windows tend to be few and small in hot arid regions, and in hot humid areas openings are large but shaded or shuttered to allow cross-ventilation and reduce light and heat penetration. In temperate regions, window areas are large or outdoor areas flow into the building, while in cold climates window openings are small and covered to reduce heat loss. But few regions are constantly the same throughout the seasons, so that seasonal variations will demand a building that responds to two or more climate types. It is already complicated enough to try to incorporate climate considerations in the heating and cooling of a building; how can daylighting considerations be included too? And how can the leap be made from certain isolated examples to applications in the building types where daylighting is likely to affect building energy performance? There are no easy answers.

DAYLIGHTING ISSUES IN LARGE BUILDINGS

The large field of energy conservation in building design is breaking down into ever more subtle categories. In residential design the classic definition of "energy conservation" is considered to be keeping negative environmental impacts out of the house. However, true energy-conscious design which is climatically responsive encourages sunshine and other positive environmental influences to enter. Lighting does not play a significant role in residential energy consumption patterns, but is a prime contributor to energy consumption in office buildings, schools, commercial low-rise buildings, health-care facilities, and warehouses. As interest in energy-conscious building design grows to include medium and large-size structures such as these, lighting and daylighting become important considerations in the overall energy balance of the building. The lighting load can account for as much as half of the total energy loads in an office building, depending on building design and location. In addition, each watt of electric lighting requires 1/2 to 1 watt of mechanical cooling to counteract its heat generation. Mechanical equipment and distribution systems are still a necessity in most of these building types—or are they? Will energy conserving design efforts in large buildings be limited to mechanical "fixes" only or will they include site and climate-responsive considerations such as passive solar heating and daylighting? Can we achieve energy independence to the extent realized in the Wallasey school? Or is a more realistic goal the reduction of mechanical equipment loads? In any case, daylighting is an integral part of the balance equation, both for its ability to eliminate large parts of the electric lighting load and also for its intimate association with heat delivery and heat loss through glazing.

The issues that are important to fenestration design in large buildings are not all concerned exclusively with light. The window, clerestory, skylight, or other aperture which admits light is also performing other functions, such as heat gain or heat loss, which affect both the occupants and the building operation. All these factors must be considered. The effect of changing any one factor in the iterative process of design must be evaluated with regard to its effect on the total building design.

The first consideration in designing for daylight is occupancy of a building: who and when? It is important to determine if people inhabiting the building have special visual requirements that dictate certain lighting qualities or illumination levels (low or high), or if they will be restrained in their movement (in a hospital bed, at a desk all day, etc.) so that exposure to direct sunlight or harsh sky glare would cause discomfort. "When" the building is occupied determines the usefulness of daylight contribution, with major daytime occupancy providing the greatest opportunity for daylight utilization. Schools and offices are prime examples since their operating hours coincide with daylight hours. Morning warm-up is a major heating load, which east-facing glazing can help supply. Afternoon overheating through west glazing should be avoided. Some health care facilities, on the other hand, are operating day and night. In this case, daylight can reduce a major electrical load but cannot totally substitute for it. Site location and availability of daylight also enter into the picture: there are not many hours of winter daylight in the northern parts of Alaska. One of the large-scale results of effective utilization of daylight occurs in the summer, however, when daylight is plentiful everywhere, and that is reduction of peak power demand on utilities. In some locales this energy-saving potential might actually reduce the demand for constructing new electrical generating facilities.

In the schematic design phase, consideration of occupancy patterns and lighting requirements will help determine the general configuration of the building and the location of various functions. An overall building configuration which is divided into wings rather than lumped together provides more daylight accessibility and, if properly oriented, more southerly exposure. (Fig. 1) As another example, attention to the lighting requirements of various staff members could lead to locating personnel with visually demanding jobs near daylight sources. Executives whose major office tasks are dictating, talking on the telephone, and holding conferences might be located in offices somewhat distant from the windows, while secretaries and clerks with predominantly visual tasks might have stations near the windows. The architect's imagination might then be stretched to find another status symbol to compensate for the executive's loss of view!

Another example of energy-conserving building layout is grouping together functions which are on the same time schedule, so that the lighting and ventilation systems for that entire section can be shut down when it is not operating. This capability assumes an appropriate control system. No matter how few watts per square foot a lighting system uses, it does not conserve energy if sixty fixtures must be switched on to provide light for only one worker.

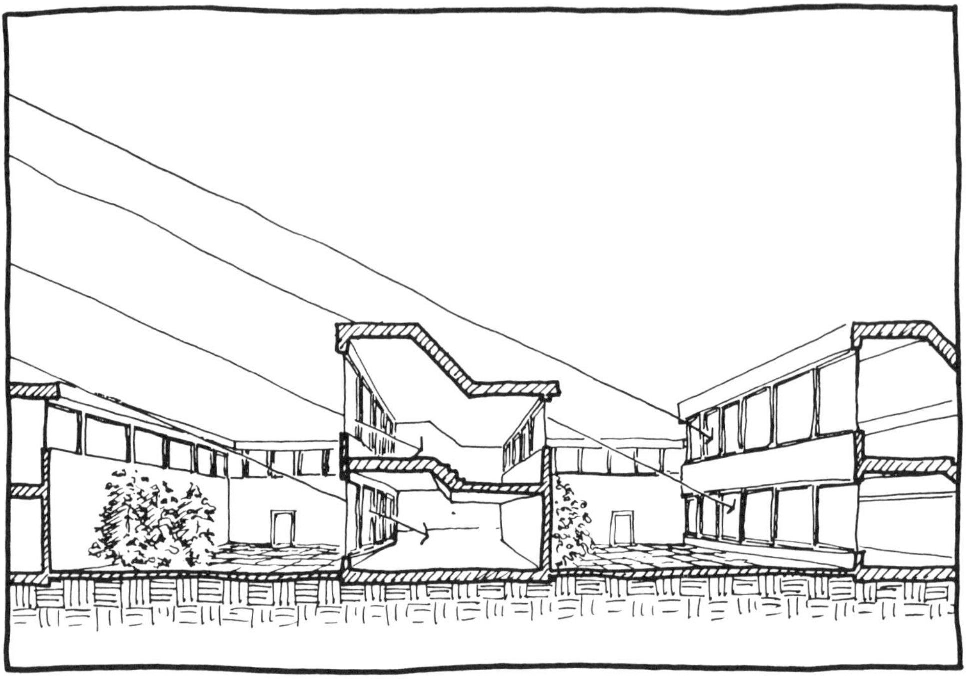

Figure 1. Wings of Light.

Five basic lighting issues must be addressed in developing the interplay among the interdependent factors that creates a successful passive solar building. These issues are: 1) achieving lighting quality which suits the uses to which each space is put; 2) designing daylight apertures which also fulfill biological and psychological desires of occupants, i.e., for view and for a psychological connection to the outside; 3) coordination of apertures for daylight with apertures for natural ventilation; 4) inclusion of sun-control mechanisms; and 5) provision of adequate levels of daylight illumination to satisfy a reasonable portion of lighting requirements and supplementing that with well-integrated electric lighting.

In coordinating the daylight design so that it complements the passive solar design approach, all the given physical parameters of the climate and the site must be considered. The weather data must be carefully evaluated and availability of daylight determined as fully as possible. This task is difficult to accomplish now due to the lack of adequate long-term measurements of daylight in different climate regions, but work is currently under way to derive conversion factors to compute available outdoor illumination data from readily available radiation data.[4] Until information of this type is available, a good general source for overcast sky daylight availability during

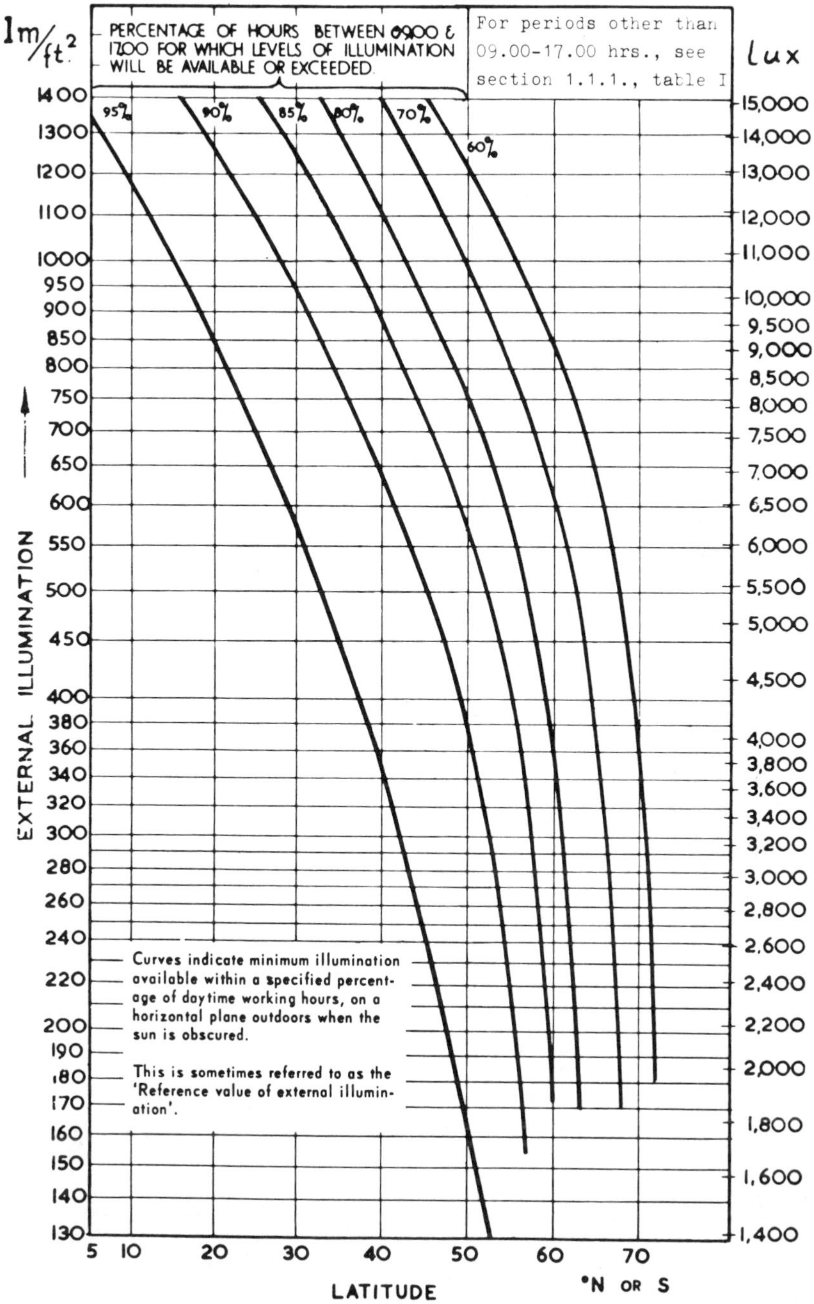

Figure 2. Availability of daylight, overcast sky. (Reference 5.)

normal working hours (9-5) is contained in a C.I.E. publication and is reproduced in Figure 2.[5] For the clear sky, illumination is related to the altitude of the sun as shown in Figure 3. Local weather station statistics on percent of sun and percent of cloud cover, coupled with an understanding of the dominant weather patterns obtained from living in a place, provide an adequate base. It appears obvious, but just as a passive solar building which works well in Sante Fe won't work well in Seattle, daylighting techniques designed for Sante Fe may be inappropriate in Seattle. In fact, the most interesting point about Pacific Northwest weather patterns is their extreme unpredictability, requiring passive solar and daylighting systems which can respond to clear sunny periods and also to long overcast spells.

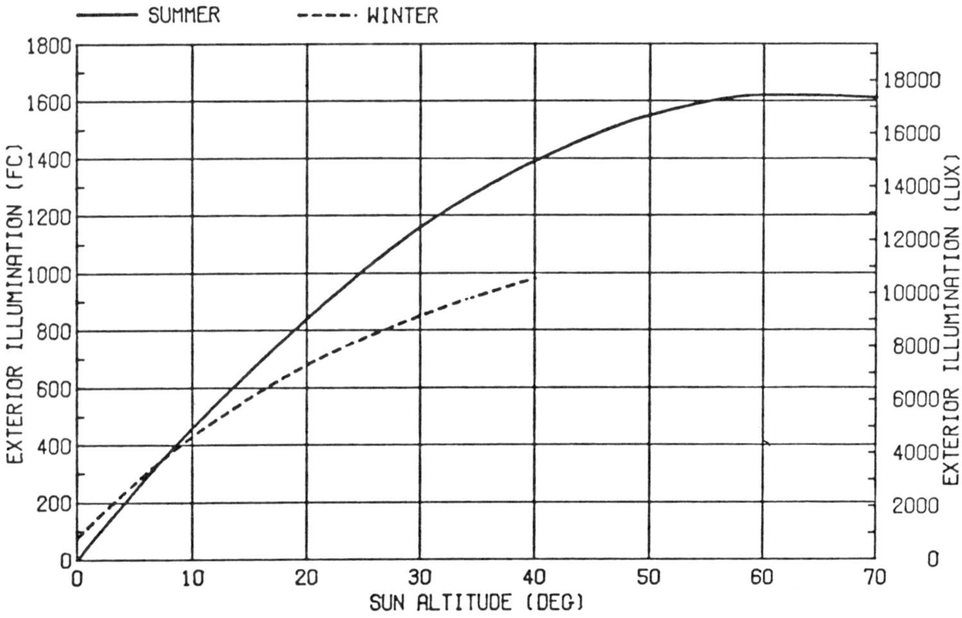

Figure 3. Clear Sky Illumination Values Relative to Sun Altitude.

In considering daylighting along with passive solar design approaches, it is important not to focus on "devices," such as a Trombe wall or a greenhouse, but rather to keep one's mind set on *merging the building form with local environmental forces* in order to minimize consumption of nonrenewable energy sources to satisfy comfort requirements of the inhabitants. Three issues must be considered: 1) solar control (allowing or preventing heat gain); 2) thermal balance (allowing or preventing heat loss); and 3) ventilation.

The issue of solar control requires an evaluation of the role of the glazing as a heat and light transmitter. Heat gain control most often involves exterior shading devices for maximum effectiveness. The shading technique that allows daylight to be utilized without adverse heat gain is a combination of clear glazing with an exterior shading device, such as an open-work horizontal overhang (Fig. 4). Deciduous trees are excellent shading devices, providing shade in summer and access to the sun in winter. (Fig. 5) The most complete and useful catalogue and workbook of shading devices is a recent reprint of the Olgyays' work at Princeton.[6]

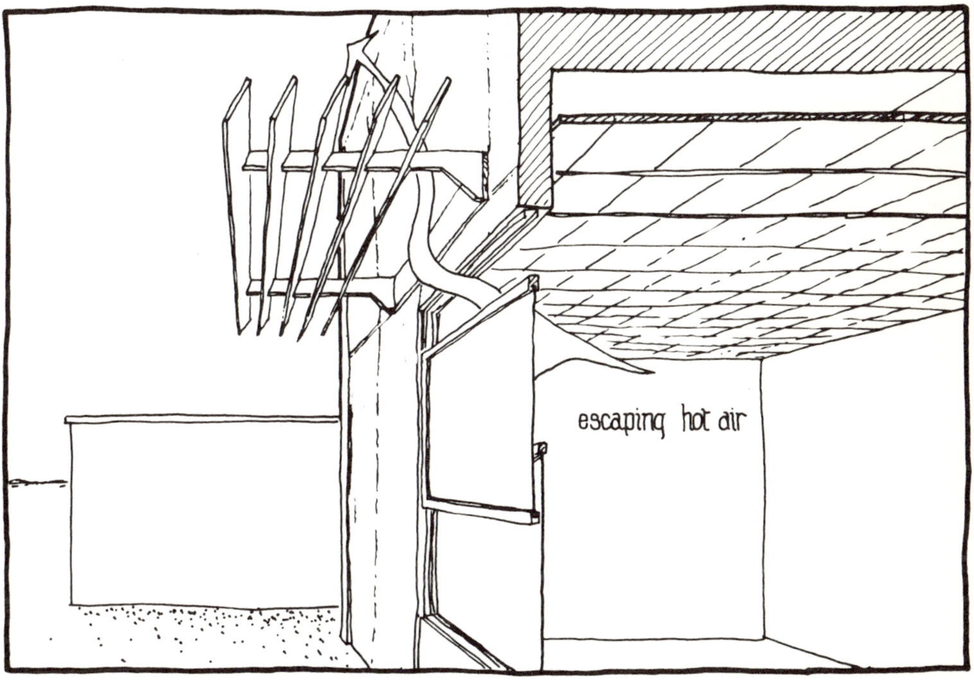

Figure 4. Example of Design Response to Convergent Issues of Using Daylight, Preventing Heat Gain, and Allowing Ventilation.

Provision of winter solar heat gain requires areas of south glazing which admit both heat and light. Clear weather brings direct sunlight into the space in winter with its concomitant problems of high contrast and glare. Even under overcast skies large areas of glazing can create glare conditions. However, openings can be placed so that patches of bright sky are removed from the direct visual path of most occupants by the use of carefully designed skylights and clerestories. Several successful passive solar buildings demonstrate this approach, for example, Ed Mazria's Maxamillian's Restaurant.

Figure 5. Deciduous Trees for Summer Shading and Winter Daylighting Effects.

Areas adjacent to the building affect the entry of light into rooms through windows as well as creating microclimate effects at the building perimeter. (Fig. 6) These effects can be used to advantage, for instance reflecting sunlight off a south-facing white wall into north-facing windows and creating a pleasant courtyard too. Reflectances of interior room surfaces play a role in the overall energy balance of a building in a different way. Room surfaces with high reflectances (70% or 80%) provide maximum interreflections, thereby increasing the light level in a room and reducing contrast between interior surfaces and the sky outside. (Fig. 7) Although this effect is generally well-known its impact on life-cycle cost analysis is striking. A two-person dormitory bedroom studied with ceiling/wall/floor reflectances of 40%/20%/10% used 4.5 times the amount of energy annually as one with reflectances of 85%/80%/50%.[7] Light interior surfaces also reflect incoming solar radiation and spread it throughout the room for absorption into the necessary storage mass.

Solar control is an important issue in relation to both thermal and also visual comfort of the inhabitants. If glazing is required at eye level, for example, for a view or because the roof is not directly above the room,

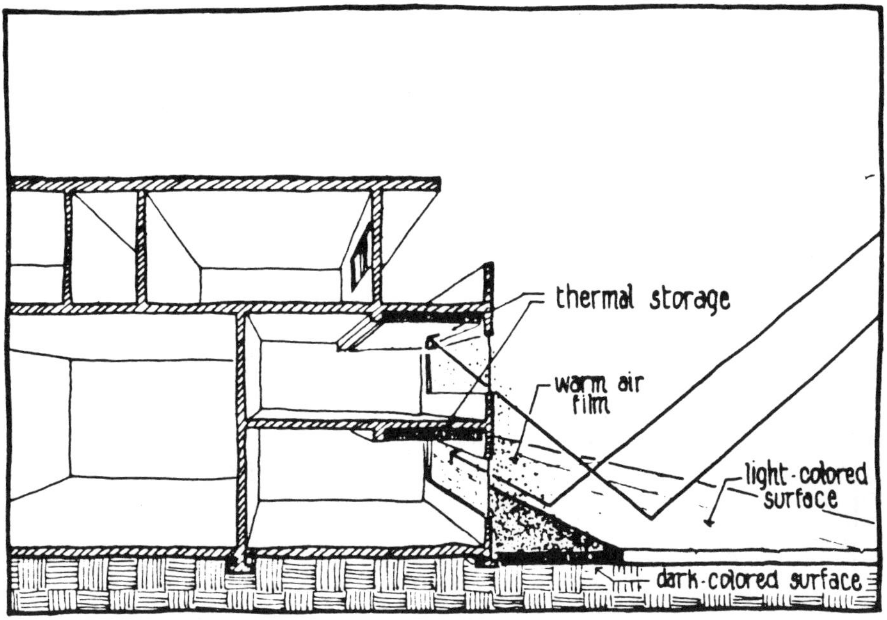

Figure 6a. Light Colored Exterior Surfaces Used to Provide Microclimate at the Building Surface.

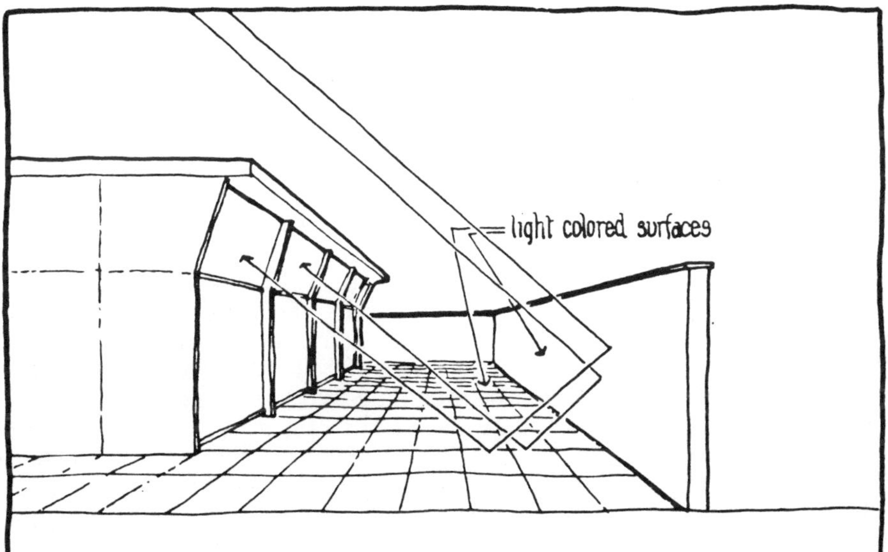

Figure 6b. Northerly Windows can Gain Daylighting From Light Exterior Surfaces.

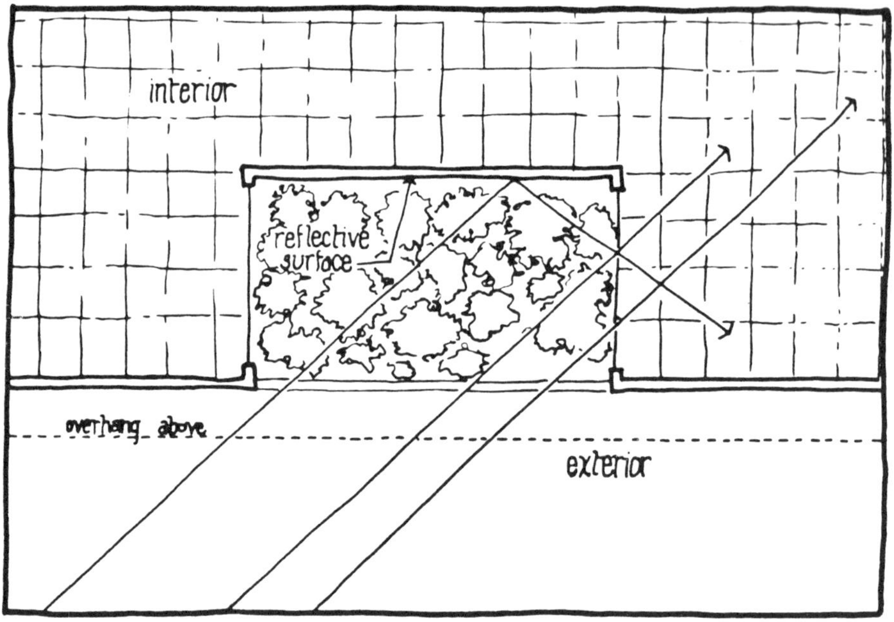

Figure 6c. Easterly or Westerly Surfaces May Utilize Reflective Surfaces for Daylighting Without Excessive Solar Gain.

either the heated space should be one where people can move freely to avoid uncomfortable conditions or some light-filtering materials should be provided for visual comfort. Some forms of glare can also be mitigated by allowing light to enter the room from at least one other side. In this case, light from one aperture will strike the surface in which the other opening is located, brightening it and reducing contrast with the sky outside.

In the category of thermal balance, which is here defined as allowing or preventing heat loss, we enter the area of window construction and management, discussed in the accompanying papers. Good window construction prevents infiltration and good window management minimizes heat loss by, among other things, providing and using night insulation. When considering daytime balance between heat transfer and daylight admission, north glazing may provide helpful thermal transfer as well as high quality diffuse light for buildings where the dominant mechanical load is cooling, such as office buildings in temperate climates. The location of north glazing with regard to proximity to people is critical, however, since radiant heat loss from the body would be increased. These considerations might lead to north clerestory glazing in a corridor, as one isolated example. To carry this example further, clear glazed transoms could then be used to admit "borrowed light" into inner working spaces. (Fig. 8) An opposite example is rooms where comfort levels are critical, such as hospital patient rooms. Both

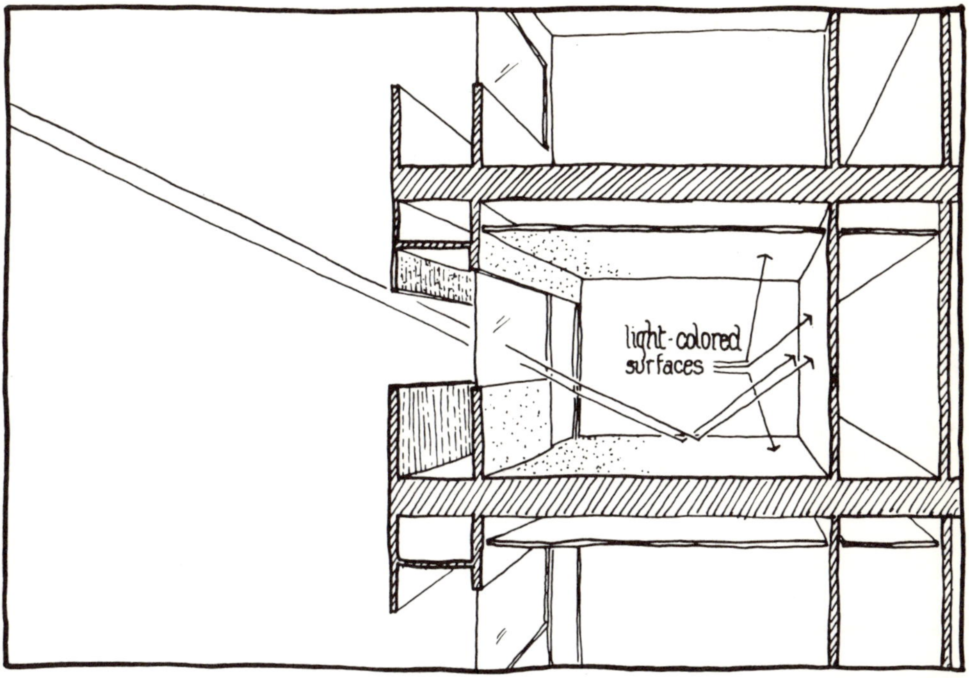

Figure 7. Light Colored Interiors Enhance Daylighting Effects.

Figure 8. An Example of "Borrowed Light".

heat transfer and daylight admission must be carefully designed to avoid overheating or glare for the usually immobile patient. A high south-facing window with a deep sill, in combination with a low shaded window for view, can solve the daylighting problem and work as a direct-gain collector too.

The third category, ventilation, is usually considered along with windows since they most often provide the ventilating openings. Alternatives are possible, however, such as doors or hinged solid panels. In using the latter, the dynamic character of both light and ventilation are reinforced by the lighting changes that take place when the solid panels are opened, as in LeCorbusier's Carpenter Center at Harvard. When considering windows for both daylighting and ventilating purposes, the effects of each must be weighed. Here it would seem that placement of windows for cross ventilation should receive priority, and that treatment of windows for daylight admittance and control should follow. With openings in opposite walls, good overall air circulation is achieved if the walls are slightly askew to the direction of the wind. (Fig. 9) In this case, windows on opposite sides will provide bilateral lighting, which can be varied by window height location. Bilateral lighting reduces contrast and makes a naturally balanced

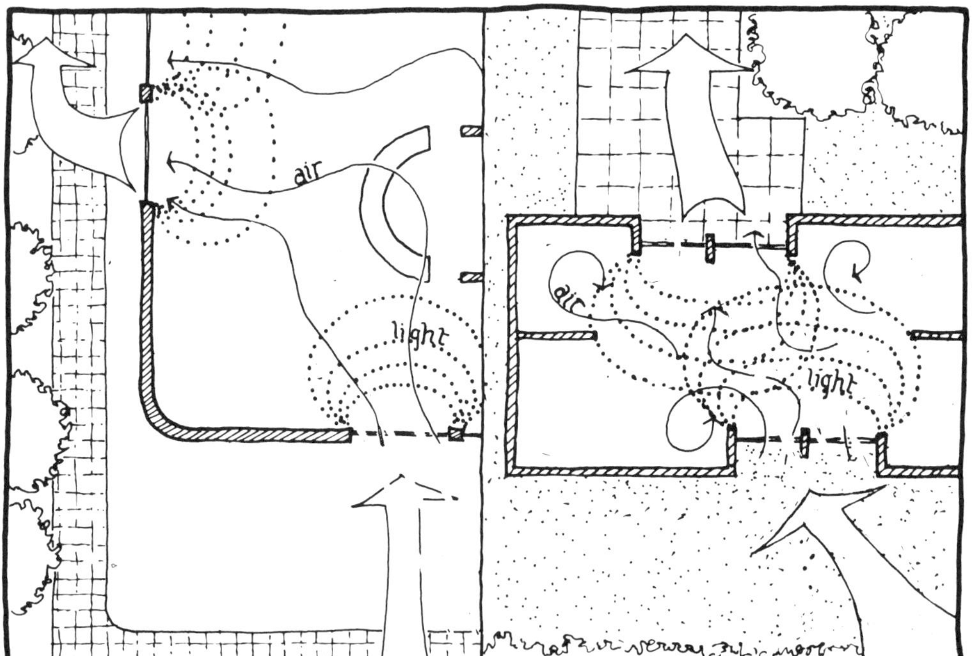

Figure 9. The Effect of Window Placement on Light Distribution and Ventilation Patterns.

luminous environment. On the other hand, if prevailing winds during the cooling season are perpendicular to the building wall, then windows in adjacent walls provide good overall circulation and minimize stagnant air pockets.[8] Light from windows in adjacent walls also reduces contrast and provides balance in comparison to a unilateral daylighting scheme. Variations in window shape, placement and treatment can also add drama.

Although we cannot forget the qualitative analysis of lighting,[9] the final quantitative evaluation of a passive or energy-conserving design proposal will be the bottom line: energy expenditure, in Btu per square foot per year or watt/hours per square foot per year; and the associated costs. Whether each window or skylight is a net energy gainer or loser becomes lost in the final equation. Certain efficiencies in one area, such as minimizing heat loss through glass in winter, may be sacrificed in order to achieve efficiency in lighting energy use. It is difficult to assess the exact trade-offs as each element of a design is altered, but preliminary results from computer simulations of thermal transfer in buildings indicate that smaller windows are not necessarily better, that the inclusion of daylight use indicates larger window areas are beneficial when they are properly oriented and insulated.[10,11,12] The continuation of these studies will give us valuable information on the effects of alternate design approaches. But it must be stressed that design to a number criteria is not the essence of good design. The essence of good design is a delightful building which requires minimal consumption of nonrenewable resources in order for the occupants to be comfortable enough to enjoy its delights.

ACKNOWLEDGMENTS

I extend my sincere appreciation to the people with whom I have been working over the last three years, especially Jim Bedrick, Ann Airy, Guy Spencer, Sue Slatkin, and Brad Andersen. Contributions from them all are visible here.

REFERENCES

1. Guirgola, Romaldo and J. Mehta. *Louis I. Kahn.* Artemis Verlag fur Architektur, Zurich and Munich, 1975, p. 187.

2. LeCorbusier, trans. by Frederick Etchells, *Towards a New Architecture*, Architectural Press, London, 1927.

3. Louis I. Kahn, Oeuvres 1963-1969, *L'architecture d'aujourdhui*, No. 142, February-March, 1969, p. 15.

4. Selkowitz, Stephen E. "Daylighting and Passive Solar Buildings." *Proceedings of the Third National Passive Solar Conference*, San Jose, California, January, 1979, pp. 271-281.

5. C.I.E. Committee E-3.2. *International Recommendations for the Calculation of Natural Daylight.* C.I.E. Publication, No. 16, Paris, 1970.

6. Olgyay, A., and V. Olgyay. *Solar Control and Shading Devices*, Princeton University Press, Princeton, N.J., 1976.

7. Spencer, Guy S. "A Method for Determining the Effects of Lighting Design Decisions on the Life Cycle Cost of Buildings." Master's Thesis, Department of Architecture, University of Washington, 1979.

8. Hastings, S. Robert and R. W. Crenshaw. *Window Design Strategies to Conserve Energy*, NBS BSS 104, National Bureau of Standards, Washington, D.C., June 1977.

9. Lam, William, M.C. *Perception and Lighting as Formgivers for Architecture*, McGraw-Hill, New York, 1977.

10. Arumi, Francisco, "Daylighting as a Factor in Optimizing the Energy Performance of Buildings," *Energy and Buildings*, Vol. 1, No. 2, pp. 175-182, 1977.

11. Bedrick, James R., A.F. Emergy, D.R. Heerwagen, C.J. Kippenhan, M.S. Millet, G.B. Varey. "The Coupling of Thermal and Lighting Simulation Programs for Evaluating the Effectiveness of Daylighting Design." *Proceedings of the Fourth National Passive Solar Conference*, Kansas City, Missouri, October, 1979.

12. Built Environment Study Teaching and Research Group. *The Incorporation of Energy Conservation Principles Into the Design of State Buildings.* Interim Report. Institute for Environmental Studies, University of Washington, Seattle, 1977.

Over, Under, Sideways, Down

An introduction to the underground section

Ken Labs

If, in 1974, you were to tell a friend or neighbor that you were going to build an underground dwelling, you might have been told that the Cold War had been over for some time. Or, perhaps, that the chipmunks and squirrels would get along just fine without your help. Today, you might receive a quizzical look, but you'd probably be pried for more information.

Interest in the "underground alternative" has expanded rapidly in the past few years, owing largely to the energy conserving potential of underground construction. A review of the more salient events of recent years provides a suitable introduction to the subject of underground building.

Prior to 1974, the literature of underground design consisted of a few promotional articles, some recommendations in *House Beautiful's* 1949-1951 "Climate Control Project," and some scattered reviews of current activity, such as *Progressive Architecture's* April 1967 issue, "The Earth." Discussion of heat transfer from underground spaces was limited to basements and fallout shelters, the latter of which addressed conditions scarcely applicable to the day-to-day use of buildings of normal occupancy. Of the former, only three papers were generally known (see references). One of these was based on experimental data, with no description of theory, while the remaining two described analytical methods, but with no correlation to observations (in fact, however, observed heating season data published in 1942 shows fairly good agreement with the later theoretical methods based on assumptions of steady state heat transfer). The main disadvantage of the empirical method has been its limited applicability to similar ground climates. The failure of steady state theory, although seemingly adequate to predict maximum winter heat loss, is its inability to model annual energy flux. The contributions of Ernst Eckert, Lyle Feisel and the continuing observations of Gordon Moore contained in this volume, therefore, represent important advances in our understanding of and ability to quantify heat heat transfer from underground spaces.

The period 1974-1975 may be viewed in retrospect as a kind of watershed in underground activity. During this time, the energy conserving potential of underground space was first scrutinized and it was concluded that "in no way can improved insulation for above ground construction begin to compete with underground buildings in the conservation of energy" (Bligh and Hamburger). Critical assesments of the role of underground design were being made in some schools, and in July of 1975, the first national conference addressing the subject "The Use of Earth Covered Buildings" was held with the support of the National Science Foundation. Shortly thereafter, largely through the efforts of the Department of Civil and Mineral Engineering of the University of Minnesota, the American Underground Space Association (AUA) was founded and representing it appeared the journal *Underground Space* (Pergamon Press).

Underground Space, an interdisciplinary journal, devoted its entire fourth issue (August 1977) to earth sheltered architecture, and it periodically carries features on building design and energy use. *Earth Shelter Digest and Energy Report* is a popular press bimonthly magazine which emphasizes earth sheltered housing and attendant issues of codes and financing. The architectural press has been slow to respond, with the noteworthy exception of two recent feature articles in the *AIA Journal* (April and November, 1978).

Underground housing has been thoroughly discussed in a definitive 300 page book, *Earth Sheltered Housing Design*, prepared by the Underground Space Center of the University of Minnesota and released in early 1978. The Underground Space Center itself was created in 1977 with an appropriation from the Minnesota Legislature; it is an active research and informational center that concentrates on design and research related to underground buildings.

A second major professional conference was held in the spring of 1978 under the title "Alternatives to Habitat: The Use of Earth Covered Settlements." It, like the 1975 conference, was organized by Frank L. Moreland, and was supported by the Department of Energy. The two volume proceedings are available from the National Technical Information Service. Short bibliographies on underground construction are distributed by the American Institute of Architects and the National Solar Heating and Cooling Information Center. An extensive and regularly updated bibliography is available from the Architecture Extension Office of Oklahoma State University.

The peculiar newsworthiness of living underground has garnered attention from the media, and the digging has begun in earnest in the past two years. Oddly enough, when *Earth Sheltered Housing Design* was being prepared in 1977, the few known contemporary examples of underground houses were designed by architects. Several of these, in fact, were built with the architect themselves as clients, either as resident or on speculation. From present appearances, earth sheltered construction has rapidly shifted to a grass-roots movement, being nurtured largely through local seminars sponsored by university extension services, alternative energy associations, and a few enterprising builders.

The professional literature has not kept pace with the popular interest. This is easily understood since it is the low technology and immediateness of underground construction that gives it its popular appeal. There seems to be, moreover, a sentiment that because no specialized hardware is necessary, professional services are dispensable. Recorded failures, however (structural and thermal), attest to the possible ramifications of too fervent a belief in "just doing it," and doing it without professional consultation. Many of the difficulties experienced by owner-builders involve problems in detailing that could be averted with architectural or engineering consultation. These incidents establish the need for an integrated approach to underground design and construction techniques.

There is, therefore, a theme tying together the seemingly diverse papers in this collection. It is a holistic theme of relating successful construction practice and the experience of builders with state-of-the-art presentations of new literature that furthers our understanding of energy transfer in the ground. It is hoped that this dual emphasis will serve the readers of the volume—and the subsequent doers—to best advantage.

SELECTED REFERENCES

(1) Houghton, F. C., S. I. Taimuty, C. Gutberlet and C. J. Brown, "Heat Loss Through Basement Walls and Floors," ASHVE *Transactions*, No. 1213, Vol. 48, 1942.

(2) Elliot, J. M. and M. Baker, "Heat Loss from a Heated Basement," ASHRAE *Transactions*, No. 1724, Vol. 66, 1960.

(3) Boileau, G. G. and J. K. Latta, *Calculation of Basement Heat Losses*, Technical Paper 292, NRC 10477, Division of Building Research, National Research Council, Ottawa, Canada, December 1968 (the method adopted in ASHRAE *1977 Fundamentals*).

(4) Bligh, T.P. and R. Hamburger, "Conservation of Energy by Use of Underground Space," in *Legal, Economic and Energy Considerations in the Use of Underground Space*, National Academy of Sciences, Washington, D. C., 1974.

(5) Moreland, F. L., editor, *Alternatives in Energy Conservation: the Use of Earth Covered Buildings*, proceedings of the conference, Ft. Worth, TX, 1975 (USGPO: SN 038-000-00286-4).

(6) *Underground Space*, Journal of the American Underground Space Association, Charles Fairhurst, Senior Editor, Dept. of Civil and Mineral Engineering, University of Minnesota, Minneapolis, MN 55455 (first issue May/June 1976).

(7) Sterling, R., editor, *Earth Sheltered Housing Design*, Underground Space Center, University of Minnesota; distributed by Van Nostrand Reinhold Co., NY, 1978.

(8) Moreland, F. L., editor, *Alternatives in Habitat: The Use of Earth Covered Settlements* proceedings of the conference, Ft. Worth, TX, 1978 (NTIS: CONF-7805138-P1 & P2).

(9) *Earth Shelter Digest and Energy Report*, WEBCO Publishing, Inc., 479 Fort Road, St. Paul, MN 55102 (first issue January/February 1979).

(10) Eckert, E. R. G., T. P. Bligh and E. Pfender, "Energy Exchange between Earth Sheltered Structures and the Surrounding Ground," *Energy*, Vol. 4, April 1979.

(11) Shipp, P. H., *The Thermal Characteristics of Large Earth Sheltered Structures*, PhD Thesis, University of Minnesota Department of Mechanical Engineering, Minneapolis, MN, 1979 (University Microfilms, Ann Arbor, MI).

(12) *Thermal Performance of the Exterior Envelope of Buildings*, DOE-ASHRAE conference, Orlando, FL, December 3-5, 1979. Several state-of-the-art papers presented dealing with heat transfer from underground spaces, implications of insulation placement and analysis methods.

(13) Jones, L. K., *Underground Construction*, bibliography prepared by LaVerne Koelsch Jones, Oklahoma State University Architecture Librarian; available from the Architecture Extension Office, OSU, Stillwater, OK 74074.

The Underground Advantage: The Climate of Soils

Kenneth Labs
Undercurrent Design Research
147 Livingston St.
New Haven, Connecticut 06511

INTRODUCTION

In considering much of the promotion of underground alternatives during the past few years, it has become increasingly clear that a number of misconceptions about the climate of soils have become so commonly popularized that they are taken for fact. A considerable body of earth temperature literature exists scattered among the disciplines of geophysics, agriculture, the biological sciences, and various branches of civil and geotechnical engineering. Unfortunately, little of this information has found its way into the world of building. The aim of this paper is to review the fundamentals of soil climate, as presented within a selected group of references. It is hoped that this effort will be of value to researchers and practitioners involved with the design of underground structures, earth pipes, and other subsurface devices, the performance of which is influenced by the thermal environment of soils.

A MATHEMATICAL MODEL

As a simplified problem in heat transfer, fluctuations in ground temperature over the course of a year can be modeled by the mathematical expression for periodic temperature variation occurring at the surface of a semi-infinite solid. The formula for this problem is rather straightforward and is developed theoretically in a number of texts on heat transfer [4, 25, 29, 60], many of which emphasize ground temperature applications.

$$T_{(x,t)} = T_m - A_s e^{-x(.00861/K)^{1/2}} \cos(.01726(t - t_o - (x/2)(.00861/K)^{1/2}$$

Where: $T_{(x,t)}$ = temperature of soil at depth x and time t
T_m = mean annual earth temperature
A_s = amplitude of surface temperature wave
x = depth below surface (ft)
K = thermal diffusivity of soil (ft^2/day)
t = time of year (days)
t_o = phase constant (days)

The model as given is valid for homogenous solids for which the assumed thermophysical values are constant over time. Rarely do these conditions truly apply to soils. Changes in weather from one year to the next can affect the values of t_o, A_s, and even T_m in the upper horizons. The value of thermal diffusivity is very sensitive to moisture content, which in itself is seasonally variable, usually reaching a minimum in the northern temperate zone during the first week of September (due largely to the effects of vegetation's "pumping" water out of the soil during summer), and a presumed maximum during the first week of March [60,20]. Moreover, soil moisture content is not necessarily uniform with depth and will be affected by soil capillarity as well as the seasonal rise and fall in ground water level.

Soils themselves are normally stratified in horizons, each of which may possess discretely different thermal properties. The most significant of these significantly differ in temperature response when compared to lower strata [20]. If the thermal characteristics of the surface layer(s) can be described, then a version of the equation for layered soils may be more appropriate. These differences may not importantly affect the validity of the model, excepting the uppermost one or two feet [5,41]. A method for predicting temperature in layered soils is developed in detail by van Wijk and Derksen [59].

In areas where ground freezing is seasonally prevalent, latent heat exchange may affect the validity of the equation. Surface freezing effects are discussed in Scott [50] and Jumikis [26]. An excellent discussion of modifications to the model to account for freezing, ground insulation, and other layering phenomena is contained in Gold and Lachenbruch [22].

Despite the many departures of earth as a medium from the idealized semi-infinite solid solution, many researchers have concluded that their field observations substantiate the general validity of the model [1, 16, 30, 34, 41, 42, 58]. It seems as though the variability of some of the "constants" tend to cancel one another, thereby negating their overall disruptive effects.

While the equation may not *exactly* conform to recorded observations at any given date, most reports indicate little more than a degree or two F error between yearly records and the "best fit" or norm of observed values, at least below the upper foot or two. Kusuda [32] and Carson [5] present different statistical techniques for determining "best fit" values for the model's variables based on recorded observations. An interesting variation of the model has been shown by Hasfurther and Burman [24] to produce good agreement between computed and recorded (daily) temperatures in the upper foot of soil, with the use of air temperature data to drive the program. A solair driven model has also been described, by Khatry *et al.*[28].

Inasmuch as the model has variables of both time and depth, subsoil temperatures may be graphed with either held constant. Holding depth constant with time variable produces the familiar sine curve, as shown in figure 1.

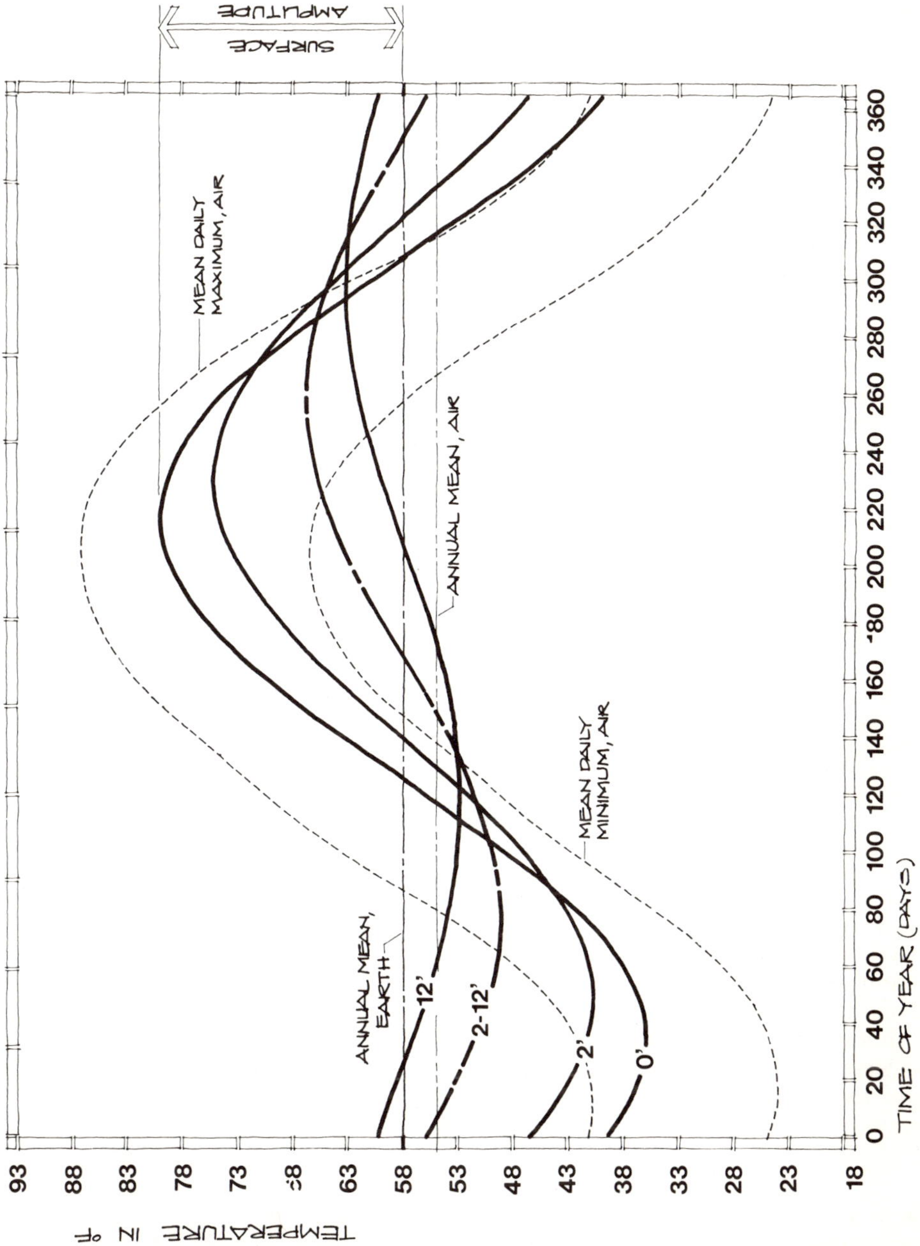

Figure 1. Ground temperatures fluctuate around the mean annual temperature over the course of a year and can be represented as a sinusoidal function related to depth. The example shown models conditions at Lexington, Kentucky, plotted against the 30 years norm of mean daily maximum and minimum air temperatures. Note the progressive lag in and flattening of the earth temperature wave with depth. Note also that earth has a higher annual mean temperature than air.

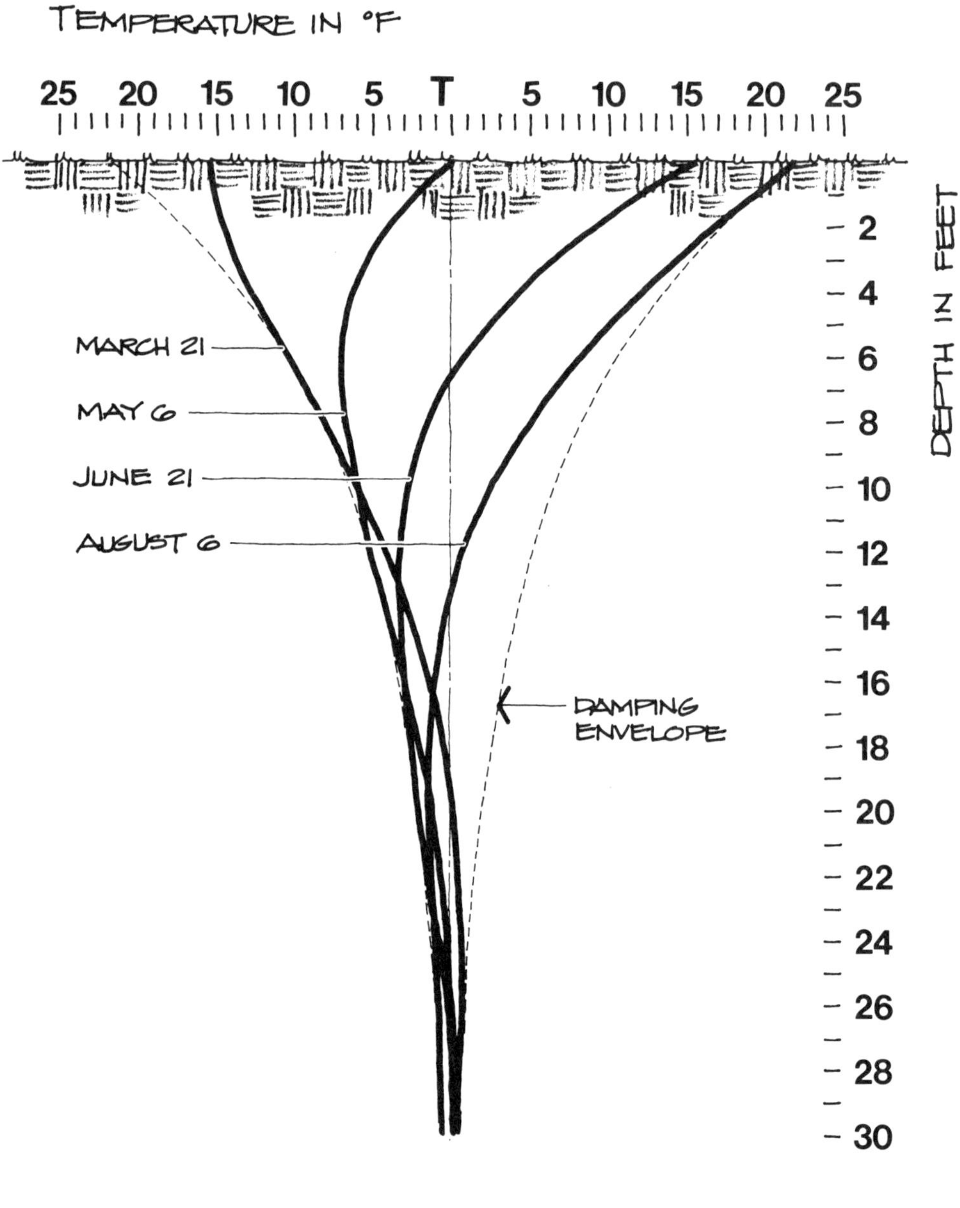

Figure 2. The four tautochrones plotted here illustrate the predicted temperature distribution in the ground at four moments during the period when surface temperature is increasing. May 6 is the approximate time at which the surface shifts from net heat loss to net heat gain.

Holding time constant and plotting temperature as a function of depth yields a graph referred to as a *tautochrone* ("same time"), depicted in figure 2. Some graphic construction techniques for the curves are explained in Jumikis[26]. Plotting several different temperature curves on each axis illustrates relationships to be discussed in the following sections.

LAG TIME

The phase of the subsurface temperature wave lags behind the surface wave in linear relation to depth (in a relatively uniform soil). The time lag is expressed numerically as:

$$L = (x/2)(116/K)^{1/2}$$

An average thermal diffusivity value for soils *in situ* has been determined by Kusuda [33] to be 0.6 ft^2/day. The computed lag time for this soil is one week per foot depth below grade.

As is frequently cited, therefore, at the depth of 12 feet in a soil of average thermal properties, maximum temperature will occur three months after that of the surface. At a depth of typical sod cover of one or two feet, the lag time is insufficient with respect to the annual cycle to be of any significant advantage. Moreover, the peak in summer near-surface temperatures has often been observed to flatten out, thereby masking and further diminishing the importance of lag time in the upper horizon. This has been explained by Carson [5] as due to increased heat-relieving daytime convection during the hot season.

AMPLITUDE

As the surface temperature wave is propagated downward through the soil, the amplitude of the temperature fluctuation decreases exponentially until it is damped out entirely. The mathematical expression for the subsurface amplitude A_x is given by:

$$A_x = A_s e^{-x(.00861/K)^{1/2}} \quad \text{or} \quad A_x = A_s e^{-xr} \quad \text{where} \quad r=(.00861/K)^{1/2}$$

Plotting the amplitude against depth yields an amplitude "envelope" or damping curve, within which the temperature curve wriggles throughout the annual cycle on its journey downward. The shape of this curve, too, is dependent on the thermal diffusivity value of the soil. Damping curves for three different soil types are plotted in figure 3.

Note that for the soil of average diffusivity, the surface temperature range is only halved at a depth of six feet, still oscillating 10 degrees above and below the annual mean. Subsurface temperatures, therefore, are not "stable" at a depth of 5 or 6 feet, as is commonly supposed, although the range is considerably less than the combined annual and diurnal range at the surface.

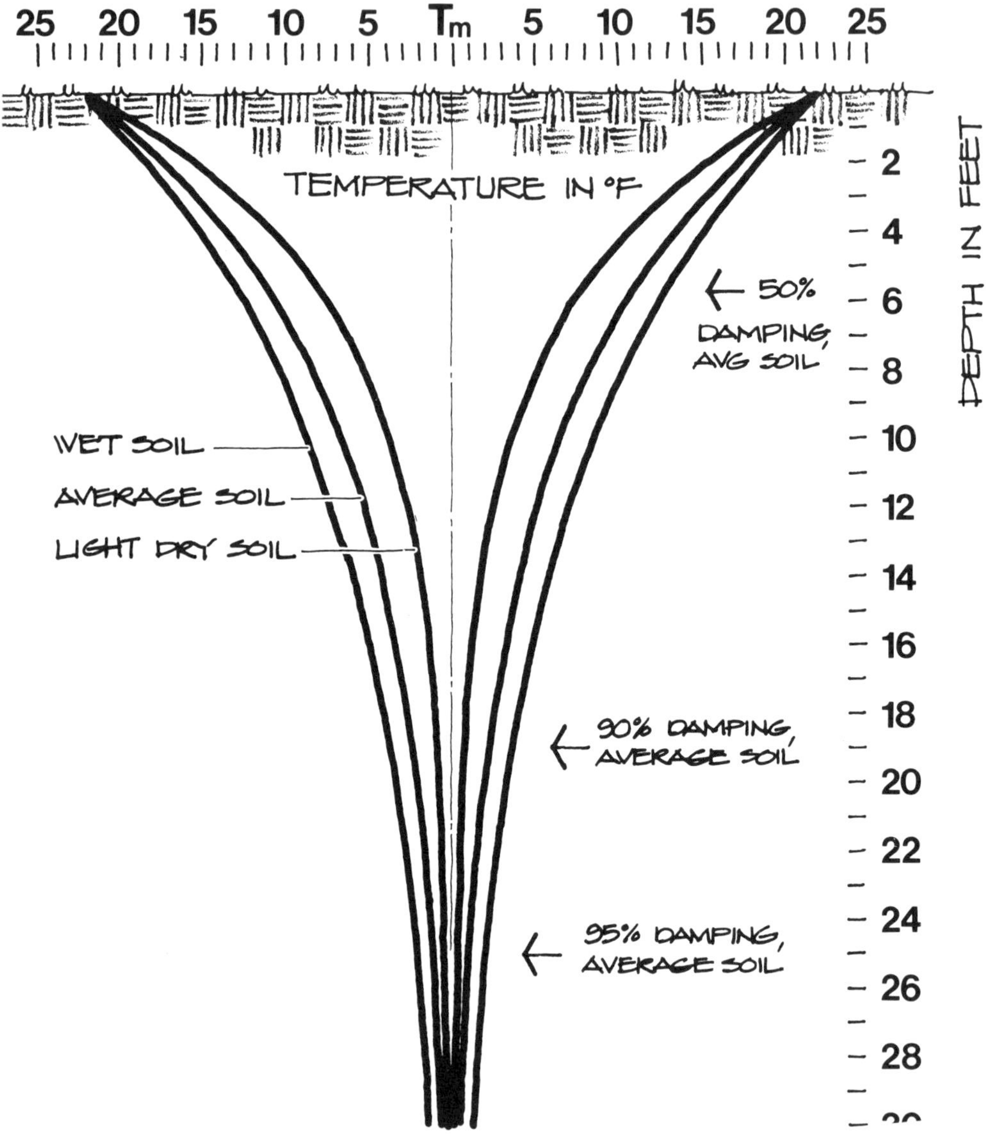

Figure 3. The amount by which temperature amplitude diminishes with depth is a function of thermal diffusivity which varies with soil type and moisture content. Note that in relatively conductive (wet) soils, the fluctuations are propagated much deeper than in less conductive (dry) soils. The graph of the maximum and minimum amplitudes with depth is referred to as the "damping curve," or "amplitude envelope."

STEADY STATE

The depth at which temperature stability occurs depends on how stable the steady state is defined to be. By rearrangement of terms in the equation, steady state depth x_{ss} can be described as:

$$X_{ss} = -[\ln(A_x/A_s)]/r$$

The expression (A_x/A_s) is the *decrement factor*. To find the depth at which the surface amplitude is reduced to 5% of the surface value, 0.05 is substituted for (A_x/A_s). This yields a depth of 25 feet for a soil of average thermal diffusivity (approximately ± 1°F). Similarly, it is found that 90% damping is achieved at a depth of 19 feet (± 2°F). The depth at which steady state occurs will also depend on ground water conditions and soil type, the influence of the latter of which may be inferred from figure 3.

VALUES OF MODEL CONSTANTS

The following generalized values are offered so that localized estimates of ground temperature may be approximated by use of the model. A discussion of influencing agents precedes the suggested substitutions.

MEAN ANNUAL GROUND TEMPERATURE

The mean annual ground temperature is determined by a host of energy exchanges at the surface, including incident solar radiation, outgoing thermal radiation, convection, and evaporation. The actual temperature of the ground is determined by the balance achieved between gains and losses. Changes brought about in the disposition of the balance of energy fluxes will result in increased or decreased energy storage, with a consequent rise or fall in ground temperature until a new equilibrium is assumed. When the equilibrium has existed for a number of years, the value of T_m will be constant throughout the soil profile with a tolerance of usually no more than ± 1 or 2°F; the uppermost layer may show greater departure from the mean, however, due to latent heat exchange [7].

GEOTHERMAL HEAT FLOW

Below the steady state depth, the geothermal gradient may be observed as an increase in T_m, ranging from ½ to 2 F per 100 feet, with 1.4 F per 100 feet as an average value. Geothermal heat production has an estimated mean value of 1.19 (microcalories) /(cm^2) (sec) or 138 Btu/(ft^2) (year) for North

America [39]. This represents about 3/10,000 of the annual solar receipt and is negligible as regards ground temperature.

SOLAR CONTRIBUTION

Latitude, aspect (slope and orientation) and albedo (reflectivity) determine the gross receipt of solar energy at the surface. As a result of greater solar loading, ground temperatures in the southern U. S. are generally higher than further north although a host of other factors preclude there from being a simple relationship between Btu's/ft^2 irradiation and ground temperature. At the local level, where other meteorological factors are similar, slopes with southerly exposures typically are found to exhibit higher ground temperatures than northerly aspects [7,64].

ELEVATION

In general, ground temperature decreases with increasing altitude, and may be characterized by a lapse rate of about 0.3 to 0.4 C per 100 meter rise in elevation [64]. On relatively gentle slopes, lower ground surface temperatures have been observed at the base of the slope than at the higher elevation of its apex. Presumably, this is attributable to the influence of the cool air pool which collects during nighttime in the basin framed by the slope and its neighbors, thereby depressing surface temperature [13]. There appears to be no data on this phenomenon as it affects deep ground temperature on rolling topography, as most slope-temperature observations have been made only near the surface, and usually only for the growing season.

SURFACE COLOR

The color or reflectivity of the surface determines how much energy is captured and available for earth heating. In agricultural applications, for example, carbon black has been added to soils to hasten the onset of the planting season [47], while whitening surface treatments have been employed to keep the surface from overheating during germination [55]. If repeated over a number of years, these manipulations could alter T_m to some degree, but they are generally found to have restricted short term effectiveness, mostly in the upper layers. Seasonal effects of some different mulching materials are compared in references [12, 23].

SOIL MOISTURE

Moisture influences, such as precipitation and evaporation from ground and plant surfaces, are responsible for much of the heat loss from the surface. Gold cites a study at Ottawa where 48% of net spring and summer solar radiation was dissipated by evapotranspiration [21]. Because of the great amounts of energy consumed in the change of phase, surface soil moisture tends to inhibit ground heating. This is evidenced in the finding that southwest slopes are characteristically warmer than south slopes; the apparent anomaly is attributed to the expenditure of morning and midday solar radiation in driving off near-surface moisture, budgeting afternoon radiation for ground heating [7, 47, 53].

The subsurface effects of rainfall or irrigation percolation are quite complex, and linked to alteration of soil thermal conductive properties as well as mass heat transfer. It has usually been observed that surface water infiltration has a cooling effect on the soil, but this may depend on season. Some researchers have expressed the belief that rain reduces maximum summer temperature but tends to raise the minimum autumn temperature [11]. Because water has a high specific heat, water-logging of the soil makes the ground more resistant to temperature change. "A wet soil is a cold soil," is an old farmer's adage expressing the slow springtime response of the soil to warming. This is confirmed in analysis of observations by Willis *et al.*, who conclude that water-laden autumn soils maintain a lower temperature during the following spring and summer than when dry [63]. Overall, individual rainfalls appear to have little immediate effect on the temperature of soil below a 6 inch depth [52], and the temporary warming of summer rain infiltration may soon be cancelled by increased surface evaporation.

IMPERMEABLE COLOR

Sealing of the surface against evaporative heat loss with impermeable cover is well known to have ground heating effect. Practical and experimental results in agricultural applications of clear and black plastic film and other vapor-restrictive mulches are well documented [12,23,47]. Gold [21] and Kusuda [34] both conducted controlled tests where black paved surfaces were compared to other ground covers, yielding much higher surface annual mean values. In Gold's experiment, the conditions were observed long enough for subsoil temperature changes to be observed at the 7 m level. Long-term temperature increases due to altered surface conditions have been modeled by Raff [46] in a paper discussing alteration of subsoil temperatures for the control of underground building climate.

WIND SHELTERING

Another means of elevating ground temperature through reduction of surface losses is to shelter the site against wind. Winter wind flow increases conduction-convection losses from within the surface, while summer windflow speeds evaporative dissipation of heat before it enters the subsoil. Windbreaks and shelterbelts, therefore, have been found to contribute to somewhat higher ground temperatures on their protected sides [49].

SNOW COVER

More important is the role of seasonally variable surface cover such as snow (and the use of wind shelter to maximize snow blanketing). The conductivity of snow has been found to vary exponentially as a function of density by a number of researchers [9]. Abel's formula $k = 0.0068W^2$, where W is snow density, is typical of these. Crawford [11] has used estimated snow cover density as it varies over the course of the Canadian winter to describe its seasonal insulating effect (see Table 1). Also using Abel's formula and a series of snow density observations from different sites across Canada, Williams and Gold [62] have calculated k values ranging between 0.04 Btu/ft(hr)F and 0.16 Btu/ft(hr)F as indicative of *in-situ* snow conditions in Canada (R-24 and R-6.4 per foot respectively). This range is typical of that found in other references [17.26] for snow conditions, although the *ASHRAE Handbook of Fundamentals* lists a conductivity of freshly fallen (7 pcf) snow as 0.34 Btu/ft(hr)F, and at 32°F as 1.3 Btu/ft(hr)F. The latter value is identical to that given for ice at the same temperature.

TABLE 1
Variation of the Average Density and Thermal Conductivity of Snow Cover*

MONTH	November	December	January	February	March	April
Average Density (pcf)	8.7	11.4	12.0	11.8	14.5	17.4
Conductivity k (Btu/ft-hr-F)	.07	.08	.09	.085	.11	.14
Resistance R (1 foot depth)	14	12.5	11	12	9	7

* From Crawford, 1952.

Recurring annual differences in snow cover between adjacent fields (drifted and open) in North Dakota were observed to produce a 3 F difference in T_m at most depths [44]. Snow cover experiments conducted by Gold show that temperature increases at a depth of 7 m can be effected by maintaining snow cover where none had previously existed [21]. Although the relationship between snow cover and ground temperature is difficult to describe with any confidence, Nicholson and Granberg [40] have found that a straight line relationship can be generalized and offer an equation for estimating deep profile ground temperatures in permafrost zones.

Seasonally-variable surface covers have been referred to as producing a "heat valve" effect that can be modeled statistically. Gilpin [17] has developed a simulation demonstrating that changes by a factor of two in the thermal characteristics of the annual surface cover can produce changes in T_m of 3.5 C on the average.

HEATED STRUCTURES

Heated buildings on the surface and buried structures also exert heating effects on the ground. A range of examples from pipelines to urban areas are discussed in references [3, 22, 26]. A method for predicting subsurface temperature within the entire field of influence of a heated surface structure is developed in a noteworthy paper by Lachenbruch [37]. Some observed temperature disturbances created by a heated underground building are discussed by Eckert *et al.* [14].

The most popular method of estimating T_m in the absence of measured data is by relationship to the mean annual air temperature, T_a. In snowless areas these two approximate each other closely: "Since the temperature of the air is chiefly dependent upon radiation and conduction from and to the ground, it can be expected that the mean annual temperatures of air and ground must be nearly the same [15]." Because the ground leads the air in temperature, T_m exceeds T_a by ½°C to 3°C in most parts of the world. In Japan, the relation $T_m = 2.5C + .96T_a$°C has been found to acceptably predict ground temperature from air temperature data [64]. Similar formulas have been offered to summarize conditions at stations in North America [20,52]. For air temperature estimates of T_m to be accurate, the air temperature station must be maintained over the same surface cover type under which T_m is being approximated [21].

GROUND WATER

In 1925, W. D. Collins published an often-referenced map which estimates ground water temperatures, based upon the allowance that mean annual ground temperature normally exceeds the mean annual air temperature by 2 to 3 F [10]. Although Collins acknowledged that this differential may exceed 5 to 6 F in "exceptional areas," it appears that no provision was made to account for the heat-trapping effects of winter snow. In areas of deep and continuous snow cover (e.g., in the northern states and Canada), T_m may exceed T_a by as much as 5 C [61]. Collins' isotherms are plotted on figure 4, on which reported T_m values for a variety of stations have been superimposed by the author.

For local site-specific estimates of T_m, temperature of well water below steady-state level provides a convenient index of deep ground temperature. The author found this an expedient method for sampling T_m in southeastern Pennsylvania. Over the course of a year, well water temperature ranged between 52.5 and 53.5°F, while ground water level rose and fell from 20 to 30 feet below grade. Mean annual temperature at a nearby (8 mi.) recording station is 53.1. In at least one experimental study, regional well water temperature showed significantly better agreement with observed ground temperature than mean annual air temperature data[57].

SURFACE TEMPERATURE AMPLITUDE

Surface temperature amplitude varies from site to site as a result of radiation factors, surface properties, and to some degree soil type. Chang [6, 7] has prepared a map depicting annual temperature range (twice amplitude) at the 4 inch depth. In general, it may be said that A_s increases with distance from the tropics, although the isopleths are not coincident with latitude. The underlying reason for the northward increase in A_s is latitudinal: although the angle between summer and winter solstice remains constant between the arctic circle and the equator, the *ratio* of summer to winter solstice altitudes increases with increase in latitude. The geographic departure from the latitudinal association may be at least partially related to moisture content: in a 1931 review of soil temperature data, Fitton and Brooks[15] observed, "the drier climates, so far as these samples go, have the larger ranges of soil temperature." Amplitude can also be expected to vary with changes in slope and aspect. Southern slopes will generally have greater amplitudes than northern slopes [15,7], owing mostly to increased heating in the summer.

SURFACE COVER

Surface cover is responsible for major differences in A_s, with well vegetated covers moderating both summer and winter extremes through shading and

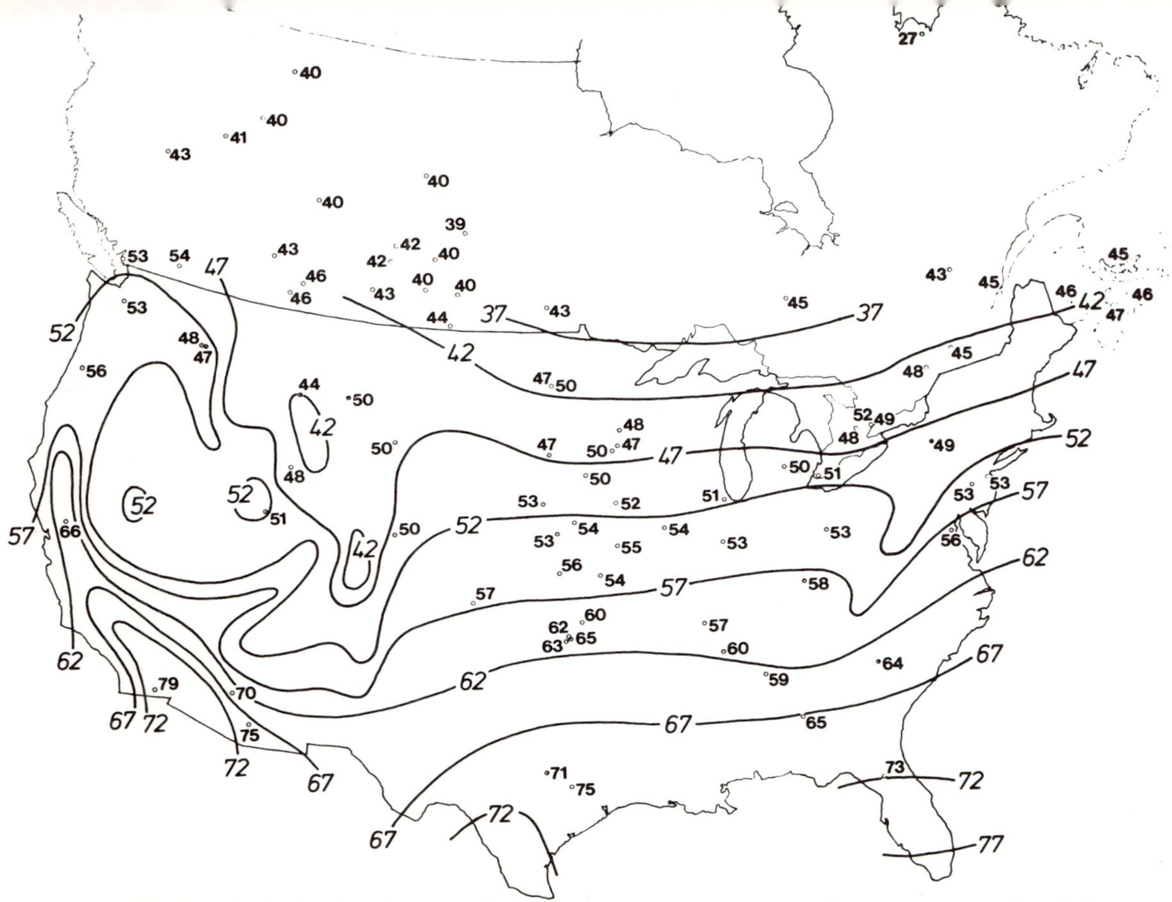

Figure 4. Collins' estimates of well water temperature isotherms are reproduced here. The author has superimposed reported mean annual earth temperatures for a variety of stations in the U.S. and Canada. The northernmost estimates by Collins are generally low, probably owing to inadequate consideration of the insulating effect of winter snow cover.

insulative effects [7, 11, 47]. Many types of mulches can be applied above the ground surface; the effect of these will be to damp out some of the temperature range before it reaches the surface. If the thermal properties of the mulch layer can be described, this damping effect can be predicted by the multi-layered soil model given by van Wijr [59, 8].

Sellers [51] cites work of deVries wherein it is suggested that the annual amplitude of a sodded surface can be assumed to exceed the annual air temperature amplitude at a 2m height by a factor of 1.25. For bare ground surfaces, a range of 1.1 to 2.0 times the air amplitude is given, with an average value of 1.3. Comparison of air temperature amplitudes with A_s values determined by least squares analysis of subsurface temperature waves projected to the surface [33] suggests that the 1.25 factor is excessive. For U. S. continental stations, projected A_s values for the most part approximate or slightly exceed one-half of the range of monthly average air temperature. Typical sod surface values for A_s fall between 19 and 26 F.

SURFACE COLOR

Surface color, of course, also affects the annual temperature amplitude where vegetation is absent. Observations by Kusuda [34] reveal greater amplitudes for a black asphalt surface than for a white painted asphalt surface. Similar observations have been made regarding different colored soils of otherwise similar thermal properties [7, 11].

CONDUCTIVE CAPACITY

Soil thermal properties determine the temperature response of the surface and undersurface to periodic heat flux. Dense, conductive soils will exhibit lower A_s than light, dry soils. The property governing this response is variously referred to as the *thermal admittance*, *conductive capacity*, or *contact coefficient*. It is the square root of the product of the conductivity and heat capacity, . Materials of high thermal admittance are good heat sinks, therefore maintain more stable surface temperature [60]. It has been pointed out that although still air has both low conductivity and low heat capacity, moving air has a high apparent conductivity. For this reason, air in motion is a much better heat sink than most soils (Table 2), so it has been said that the earth surface is in much better thermal contact with the air than with the deep ground [46,22]. Raff [46] supposes that for bare earth, the conductance to the air is 23 times as great as the conductivity through the first meter of ground. The conductive capacity is an important factor in the design of earth-air heat exchangers, although it often appears in a variation of the form in these discussions [14,25].

TABLE 2
Thermal Properties of Different Media*

Medium	Heat Capacity Btu/cf-F	Thermal Diffusivity sf/h	Thermal Conductivity Btu/h-ft-F	Conductive Capacity Btu/sf-$\sqrt{h}$ -F
New Snow	1.9	.02	.04	.25
Old Snow	14	.01	.16	1.5
Dry Sand	19	.005	.1	1.3
Wet Sand	25	.04	1.0	4.9
Sandy Clay (15% water)	37	.014	.5	4.5
Organic Soil	36	.02	.7	4.9
Wet Marshy Soil	44	.01	.5	4.7
Still Air	.019	.77	.014	.016
Stirred Air:				
very stable	.019	3.9×10^3	72.5	1.2
neutral	.019	3.9×10^5	7200	12.00
very unstable	.019	3.9×10^7	720000	120

*from Priestly. 1959 (rounded off), except for thermal conductivity

THERMAL DIFFUSIVITY

Thermal diffusivity varies with soil type as related to both density and water content, the latter of which affects both conductivity and specific heat. The relationship between K and moisture content for some sample soils is given in figure 5. Because K is intimately linked to soil moisture, it itself may be a time dependent variable, leading to an unsymmetrical waveform and altering lag time and damping depths seasonally [60].

In-situ measurements of heat flow by the University of Minnesota from an underground structure reveal that moisture content is of much greater significance than soil type itself, except inasmuch as soil type determines moisture retaining capacity.

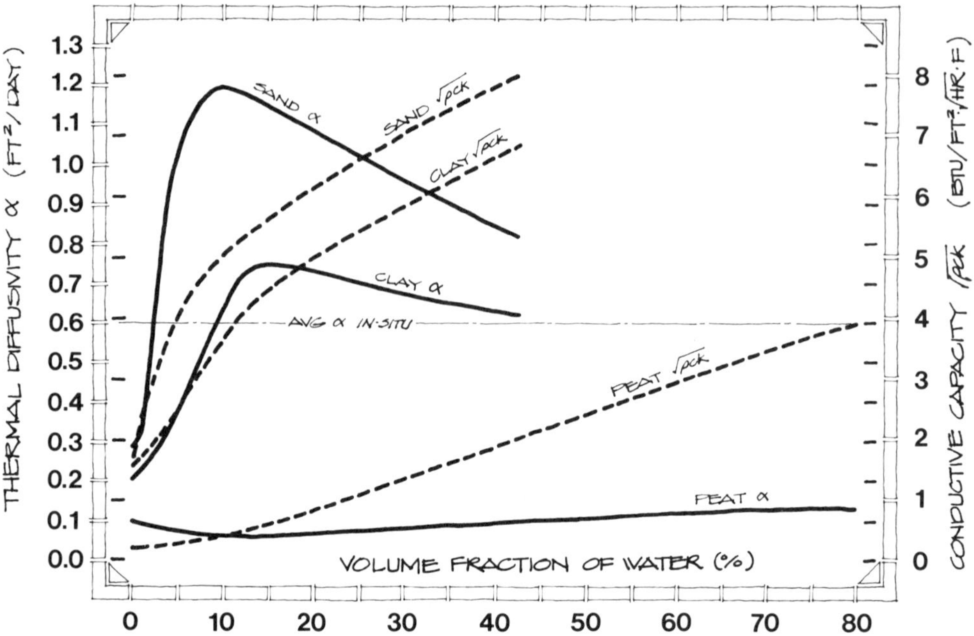

Figure 5. Thermal diffusivity and thermal admittance, or conductive capacity, are dependent on water content and soil type. The major influence is the effect of increasing soil moisture on increasing conductivity, but as water content increases, so also do density and specific heat (Graphs reproduced after van Wijk,[60])

PHASE CONSTANT

Integration of the ground temperature equation yields a sinusoidal formula that describes heat flow. Setting depth X equal to zero provides an expression for heat flow at the surface, which can be seen to lead the temperature curve by one-eight cycle, or 46 days. Given solar radiation as the driving heat source for the ground, it can be expected that maximum surface temperature will follow 46 days after the summer solstice, on August 6. Inspection of data from 29 earth stations analyzed by Kusuda [33] supports this prediction, with a tolerance range of ±10 days. Given that maximum surface temperature occurs on August 6, the value for t_o may be concluded to be 34.5 days (see figure 6).

Similar analysis of air station data reported by Kusuda reveals that maximum air temperature occurs, on average, 46 days after the solstice, with a tolerance of ± 3 days for the stations represented. Observations by Gold show that air temperature is almost in phase with the 10 cm (4") ground temperature [20]. This verifies the excellent thermal contact between air and surface. Shifts in the phase of the wave may occur as a result of heat exchange mechanisms that are not in phase with solar radiation [60]. Variable surface covers and seasonally-variable soil thermal properties have already been discussed.

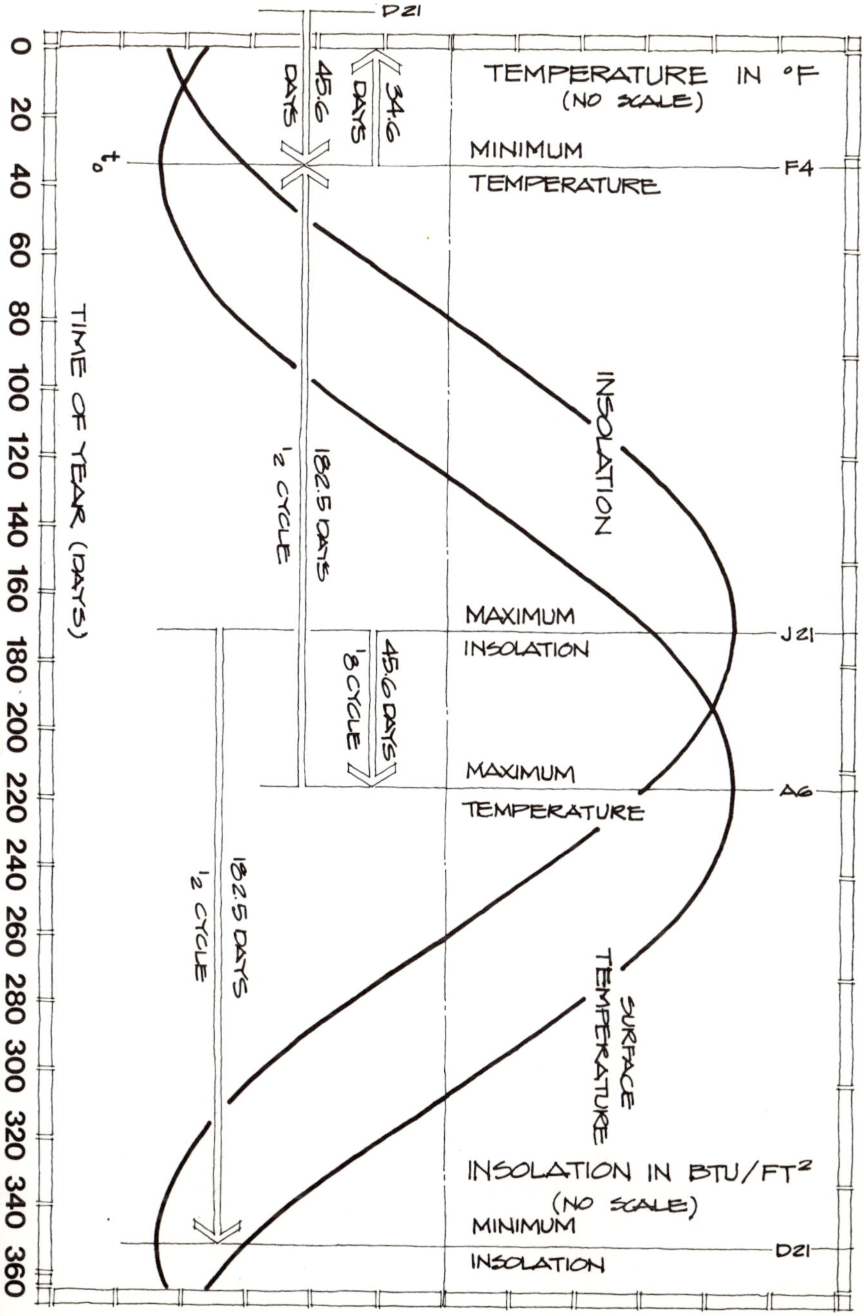

Figure 6. Surface temperature is initially driven by solar radiation. According to heat conduction theory, the surface temperature wave follows one-eighth cycle in phase behind heat flux. In theory, then, surface temperature should peak 46 days after the summer solstice, or on August 6. Least squares analysis of ground temperature observations generally substantiate this (as an idealized sine curve), although the actual recorded peak often precedes the first week in August slightly. Generally, peak temperature is coincident with the maximum mean daily average air temperature, since the ground is primarily responsible for air heating.

SURFACE SHADING

It is desirable for comfort control efficiency to retard the surface wave as much as possible, so as to maximize the phase difference between air and subsoil temperatures. It can be speculated that complete shading of the surface will effect a shift of 6½ weeks, since the shaded surface would be heated by air rather than the sun. Maximum surface temperature under a fully shaded surface, therefore, may be expected to occur 46 days after August 6, on September 21. Under a well-shaded corn crop canopy, it has been reported [44] that maximum surface and corresponding subsurface temperatures occurred 1½-2 months later than these maxima at sodded, mulched, and plowed fields. Other investigators report lesser shifts under shaded conditions, in one case as little as two weeks [54]. Little other data seems to be available to test this hypothesis.

HEAT STORAGE IN THE GROUND

A second integration of the temperature model yields an expression for heat storage, from which the total heat flux in and out of the surface during the year can be determined, utilizing the proper limits [29, 50, 51]. Penrod [43] has thus calculated annual heat storage, utilizing published observations of earth temperature to derive values for the model. He found that only 8810 Btu/ft^2 (2.27% solar) was stored in sod covered ground at Lexington, Kentucky, and 7860 Btu/ft^2 (2.55% solar) annually at Ottawa, Canada. This theory has been tested by measurements of heat flow in the ground with good agreement by Pearce and Gold [41].

Realizing that so little of the heat exchanged at the surface is transacted with the subsurface has prompted some proposals to increase the absorption and retention. Raff [46] has modelled several surface covers to "test" their effect on ground temperature. A paper by Stephenson [56] recommends winter straw mulch-insulation designed to prevent the ground from freezing. An interesting application of preventing subsurface heat from escaping has been suggested by Robinsky and Bespflug [48]: their subsurface horizontal insulation system raises the frost depth, allowing construction of shallow foundations in northern climates.

CRYPTOCLIMATE

Integration of the ground temperature equation with respect to depth, and division of this integral by the height of a selected soil profile yields an expression of the form:

$$T_{(a\text{-}b,t)} = T_m + A_s/[(a\text{-}b)r(1.414)]e^{-xr} \cos[1.0173(t\text{-}t_o\text{-}46)\text{-}xr] \Big|_b^a$$

Where a = upper bounds of profile H
b = lower bounds of profile H
$r = (.0086/K)^{1/2}$

This sinusoidal function predicts the *average* temperature of a soil profile extending from depth a to depth b. A two to twelve foot profile, for instance, may be modeled to represent the (undisturbed) average thermal conditions over the depth of placement of a fully subgrade wall. While this profile temperature is itself altered by the presence of a heated subsurface building, it can be used to characterize and compare "ambient" undersurface temperature climates.

UNDERSURFACE DEGREE DAYS (USDD)

By computing the area bounded by the 2-12' profile curve and a selected temperature index, an annual degree day value can be assigned to the given conditions. Figure 7depicts the integrated average 2-12' profile temperature prediction at Lexington, Kentucky. A selected 65°F index provides the bounds by which the area is computed, through another integration of the formula (this time with respect to t). The resultant USHDD value is found to be 2,660, comparing to an annual surface HDD norm for this location for 4,700. This represents a savings of about 43%.

Cooling degree days for the undersurface can similarly be computed by finding the area defined by the profile curve as the upper bounds, and the selected index as the lower bounds. If the upper limit of the comfort zone (78 F Effective Temperature ET) is chosen as the lower bounds, in the Lexington example the total USCDD will be zero, since the curve doesn't exceed comfort level. If the 65 F index is used for calculating USCDD, the value is a mere 91.

A regional survey by the author of undersurface heating degree days by a less sophisticated method [35] indicated that heating requirements can be completely eliminated in some southern regions, while in some others, benefits appear to be fairly negligible. Areas of particular benefit are distributed throughout the northern midwest, where savings on the order of 40% are predicted. In many milder climates where solar heating is especially effective, reduction of heating load by subsurface placement may be meaningless; moreover, it may be counterproductive if ground temperatures are high, and much of the potential for capturing summer breezes is lost.

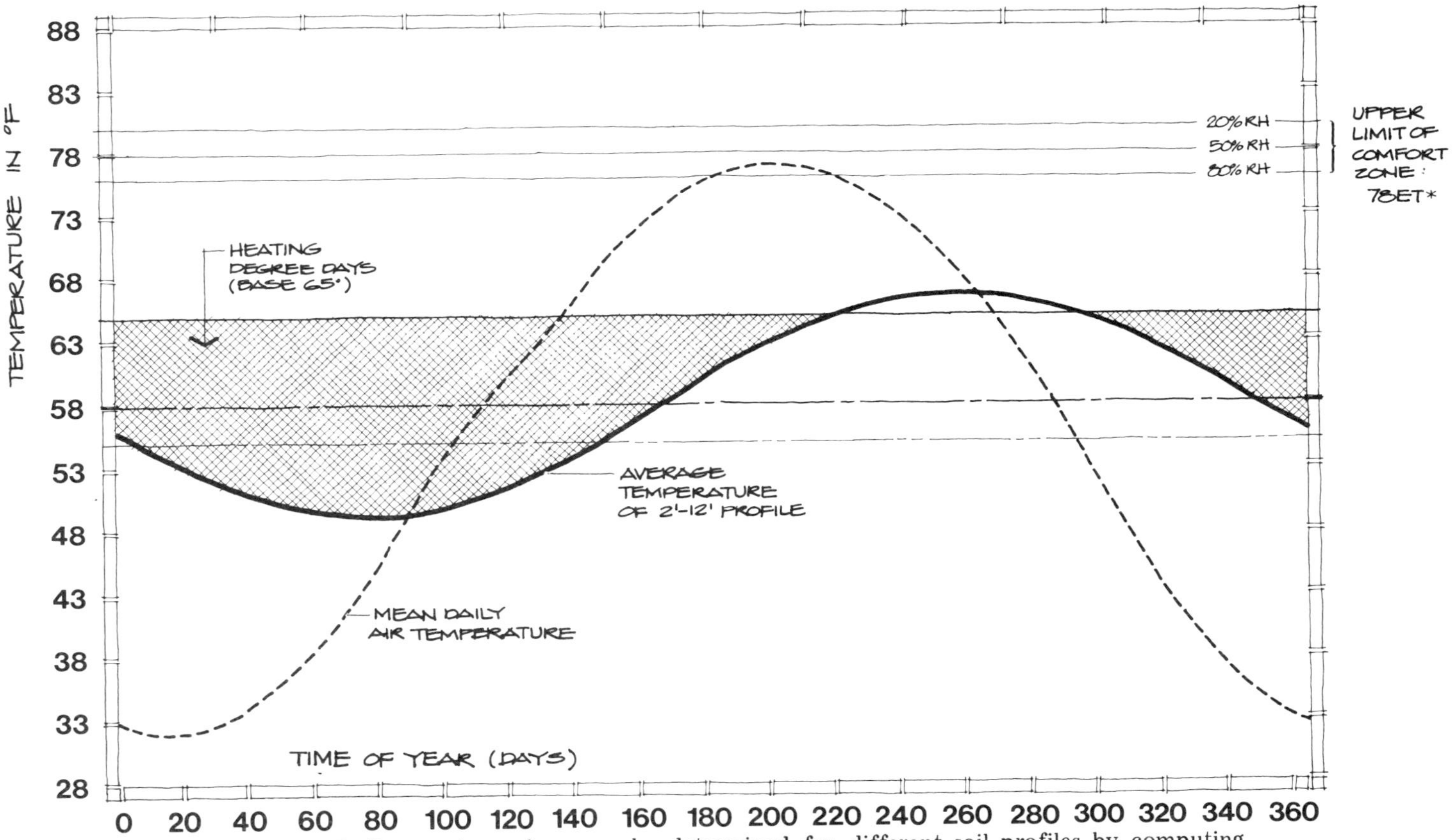

Figure 7. Degree day values can be determined for different soil profiles by computing the area above the profile's average temperature curve and a selected degree day index, here chosen as 65 F. For this 2'-12' deep profile in Lexington, Kentucky, the profile "sees" 2660 heating degree days over the course of a year, as opposed to 4720 above the surface. Note that the peak profile does not even approach the upper limits of the comfort zone (given here as 78ET*, for three different levels of relative humidity).

It is appropriate to ask how degree day values computed for the undersurface compare to the conventional variety as an indicator of heating loads or fuel consumption. As has been discussed previously, the air may be regarded as a much better heat sink than the ground, owing to convective transfer. Priestly [45] gives values of thermal diffusivity and conductive capacity for different soils and conditions of air movement. From these (table 2), it can be seen that the conductive capacity of air significantly exceeds that of soil. From the same data, the "apparent" conductivity of moving air (an abstraction) can be computed, and it, too, in its unstable condition, exceeds that of soil.

From this difference in conductive capacity, one may deduce that an identical initial temperature differential between interior and outdoor air versus exterior soil will result in different rays of heat transfer over time. As the soil conducts away heat, its own temperature increases, thereby retarding the rate of further losses. The moving atmosphere, on the other hand, continually washes heat from the surface. Consequently, one may further deduce that the value of one undersurface degree day is something less than one unit of its above-surface counterpart. Further analysis is required to quantify this relationship, although it can be expected to vary with soil type. The implication is that the realized reduction in heating demand will be considerably greater than the directly-compared savings (larger than 43%, in the Lexington example).

SUMMARY AND CONCLUSION

This paper has described observed influences on ground temperature, and has discussed these as variables in the classic equation for predicting temperature within a semi-infinite solid with periodic temperature variation at the surface. Sufficient information has been provided to enable the reader to approximate ground temperatures at specified times and depths within generalized geographic locations. Suggestions as to sources for more localized information have been given, and the effects of surface and other local factors on shifting values from generalized norms have been discussed. Numerous temperature modification techniques have been presented or may be inferred from these discussions.

Manipulation of the temperature equation itself has been shown to provide a method of calculating the average temperature of a specified soil profile. This profile average can be used to simulate thermal conditions in undisturbed earth as an index of undersurface ambient temperature climate corresponding to various building placement depths.

From the profile average temperature, a method has been presented for calculating undersurface heating and cooling degree days. An example indicates significant reductions in heating and cooling loads can be achieved through subsurface placement. Previous studies suggest that subsurface construction is far more advantageous in some geographic regions than others, and even of negative benefit in still others.

Undercurrent Design Research is presently beginning an analysis of the continental United States in a two part effort to a) evaluate regional benefits of underground construction and to b) frame the benefits and compromising implications of underground placement within the overall range of other climatic design strategies. It is hoped that this approach will help clarify the role of earth/sheltering among other passive design alternatives on a regional basis.

In general it may be said that the undersurface offers a milder, less stressful building climate than that of the surface. Reductions in annual temperature fluctuations, elimination of daily fluctuations, and the reduced rate of heat transfer to and from underground spaces have enormous energy (and first cost) savings implications for mechanical comfort control systems [36].

However, while it can be concluded that subsurface construction offers superior thermal isolation from the surface, it must also be acknowledged that the climatic amenities of solar heat and summer breezes exist only above grade. Inasmuch as design itself is a process of synthesis and integration of parts and ideas, it is all important that all alternatives be weighed against and with one another at the outset. Only then can underground building begin to be considered as architecture.

REFERENCES

1. Algren, A. C., "Ground Temperatures as Affected by Weather Conditions," *Heating, Piping and Air Conditioning*, vol. 21, no. 6, June 1949, pp. 111-116.

2. *ASHRAE 1978 Applications Handbook*, Chapter 12, "Environmental Control for Survival," American Society of Heating, Refrigertaing, and Air-Conditioning Engineers, Inc., New York, 1978.

3. Brown, W. G., "Graphical Determination of Temperature Under Heated or Cooled Areas on the Ground Surface," *Technical Paper* No. 163 of the Division of Building Research, National Research Council of Canada, Ottawa, October 1963. (NRC 7660)

4. Carslaw, H. S., J. C. Jaeger, *Conduction of Heat in Solids*, Clarendon Press, Oxford, 1959.

5. Carson, J. E., "Analysis of Soil and Air Temperatures by Fourier Techniques," *Journal of Geophysical Research*, vol. 68, no. 8, April 1963, pp. 2217-2232.

6. Chang, Jen-hu, *Ground Temperature*, vol. I, Blue Hill Meteorological Observatory (Harvard University), Milton, Massachusetts, June 1958.

8. Cochran, P. H., "Thermal Properties and Surface Temperatures of Seedbeds", Pacific Northwest Forest and Range Experiment Station, U. S. Dept. of Agriculture Forest Service, Portland, Oregon.

9. Cold Regions Research and Engineering Laboratory, "Properties of Snow," *Cold Regions Science and Engineering*, III—A1, Part III: Engineering, Section A: Snow Engineering. U. S. Army Materiel Command, Hanover, New Hampshire, December 1964.

10. Collins, W. D., "Temperature of Water Available for Industrial Use in the United States," U. S. Geological Survey, *Water Supply Paper* 520-F, 1925.

11. Crawford, C. B., "Soil Temperatures: A review of Published Records," in *Frost Action in Soils*, Highway Research Board Special Report No. 2, NAS/NRC Publication No. 213, National Research Council, Washington, D. C., 1952.

12. Davies, J. W., "Mulching Effects on Plant Climate and Yield," Technical Note No. 136, World Meteorological Organization No. 388, Geneva, 1975 (distributed by UNIPUB, New York).

13. Dreibelbis, F. R., "A Summary of Data on Soil and Air Temperatures at the North Appalachian Experimental Watershed, Coshocton, Ohio," Soil Science Society of America *Proceedings*, vol. 15, 1950, pp. 394-399.

14. Eckert, E. R. G., "The Ground Used as Energy Source, Energy Sink, or for Energy Storage," *Energy* (Oxford), v. 1, no. 3, September 1976, pp. 315-323.

15. Fitton, E. M., C. F. Brooks, "Soil Temperatures in the United States," *Monthly Weather Review*, vol. 59, no. 1, January 1931, pp. 6-16.

16. Fluker, B. J., "Soil Temperatures," *Soil Science*, vol. 86, no. 1, July 1958, pp. 35-46.

17. Gilpin, R. R., and B. K. Wong, "'Heat-valve' Effects in the Ground Thermal Regime," *Journal of Heat Transfer* (ASME), vol. 98, no. 4, November 1976, pp. 537-542.

18. Givoni, B., "Modifying the Ambient Temperature of Underground Buildings," presentation to the *Conference on Earth Covered Settlements*, Ft. Worth, Texas, May 1978. To be published in the Proceedings, Frank L. Moreland, editor, by the Division of Buildings and Community Systems, DOE.

19. Goetz, R. E., Jr., "Use of Degree Day Data to Compare Energy Consumption in Surface and Subsurface Construction," unpublished Masters thesis, University of Maryland, Department of Meteorology, 1975.

20. Gold, L. W., "Influence of the Snow Cover on the Average Annual Ground Temperature at Ottawa, Canada," Report NRCC 7505, National Research Council of Canada, Ottawa.

21. Gold, L. W., "Influence of Surface Conditions on Ground Temperature," *Canadian Journal of Earth Sciences*, vol. 4, April 1967, pp 199-208.

22. Gold, L. W. and A. H. Lachenbruch, "Thermal Conditions in Permafrost—A Review of North American Literature," North American Contribution to the *2nd International Conference on Permafrost*, Yakutsk, U.S.S.R., 1973, published by the National Academy of Sciences, Washington, D. C.

23. Hanks, R. J., S. A. Bowers, L. D. Bark, "Influence of Soil Surface Conditions on Net Radiation, Soil Temperature, and Evaporation," *Soil Science*, vol. 91, no. 4, April 196l, pp. 233-238.

24. Hasfurther, V. R., R. D. Burman, "Soil Temperature Modeling Using Air Temperature as A Driving Mechanism," *Transactions*, American Society of Agricultural Engineers, vol. 17, no. 1, January/February 1974, pp. 78-81.

25. Ingersoll, L. R., O. J. Zobel, A. C. Ingersoll, *Heat Conduction, with Engineering, Geological, and other Applications*, University of Wisconsin Press, Madison, 1954 (Revised Edition).

26. Jumikis, A. R., *Thermal Geotechnics*, Rutgers University Press, New Brunswick, N.J., 1977.

27. Kersten, M. S., "The Thermal Conductivity of Soils," Highway Research Board 28th Annual Meeting *Proceedings*, National Research Council, 1948.

28. Khatry, A. K., M. S. Sodha and M.A.S. Malik, "Periodic Variation of Ground Temperature with Depth," *Solar Energy* (Pergamon), vol. 20, no. 5, 1978, pp. 425-427.

29. Kirkham, D., *Advanced Soil Physics*, Wiley-Interscience, John Wiley and Sons, Inc., New York, 1972.

30. Kristensen, K. J., "Temperature and Heat Balance of Soil," *Oikos* (Copenhagen), vo. 10, no. 1, 1959, pp. 103-120.

31. Kuiper, L. D., "A Thermal Model for the Surface Temperature of Materials on the Earth's Surface," Technical Information Series No. 1, July 1976, Iowa Geological Survey, Iowa City, Iowa.

32. Kusuda, T., "Least Squares Technique for the Analysis of Periodic Temperatures of the Earth's Surface Region," *Journal of Research of the National Bureau of Standards*, 71C, No. 1, January/March 1967, pp. 43-50.

33. Kusuda, T. and P. R. Achenbach, "Earth Temperature and Thermal Diffusivity at Selected Stations in the United States," ASHRAE *Transactions*, Article No. 1914, 1965.

34. Kusuda, T. and B. A. Peacy, "The Effect of Ground Cover on Earth Temperature," in *The Use of Earth Covered Buildings*, F. L. Moreland, editor, National Science Foundation, Washington, D. C., 1975 (U. S. GPO #NS 1.2:EN2/8).

35. Labs, K. B., "The Architectural Use of Underground Space: Issues and Applications," Masters thesis, Washington University School of Architecture, St. Louis, 1975. (Available from the author).

36. Labs, K. C., "Earth, Energy, Architecture," in *Toward Energy Efficient Design*, Walter Wagner, editor, Architectural Record Books/McGraw-Hill Book Co., New York, 1980. (in press).

37. Lachenbruch, A. H., "Three Dimensional Heat Conduction in Permafrost Beneath Heated Buildings," U. S. Geological Survey *Bulletin* 1052-B, 1957, pp. 51-69.

38. Langbein, W. B., "Computing Soil Temperatures," *Transactions* of the American Geophysical Union, vol. 30, no. 4, August 1949, pp. 543-547.

39. Lee, W. H. K., editor, *Terrestial Heat Flow*, Geophysical Monograph series No. 8, American Geophysical Union of the NAS/NRC, publication no. 1288, Washington, D.C., 1965.

40. Nicholson, F. H. and H. B. Granberg, "Permafrost and Snowcover Relationships near Schefferville," *2nd International Conference on Permafrost.* Yakustsk, U.S.S.R., .1973 National Academy of Sciences, Washington, D.C.

41. Pearce, D. C. and L. W. Gold, "Observations of Ground Temperature and Heat Flow at Ottawa, Canada," *Journal of Geophysical Research*, Vol. 64, no. 9, September 1959, pp. 1293-1298.

42. Penrod, E. B., J. M. Elliott, W. K. Brown, "Soil Temperature Variation at Lexington, Kentucky," *Soil Science*, vol. 90, no. 5, November 1960, pp. 275-283.

43. Penrod, E. B., O. W. Stewart, "Analysis of Geophysical Data at Three Locations," *Soil Science*, vol. 104, no. 2, August 1967, pp. 86-98.

44. Potter, L. D., "Yearly Soil Temperatures in Eastern North Dakota," *Ecology*, vol. 37, no. 1, January 1956, pp. 62-70.

45. Priestly, C. H. B., *Turbulent Transfer in the Lower Atmosphere*, University of Chicago Press, Chicago, 1959.

46. Raff, S. J., "Ground Temperature Control," *Underground Space* (Pergamon) vol. 3, no. 1, July/August 1978, pp. 35-44.

47. Richards, S. J., R. M. Hagan, T. M. McCalla, "Soil Temperature and Plant Growth," in *Soil Physical Conditions and Plant Growth*, vol. II of Agronomy: A Series of Monographs, B. T. Shaw, editor, Academic Press, New York, 1952.

48. Robinsky, E. I., K. E. Bespflug, "Design of Insulated Foundations," *Journal of the Soil Mechanics and Foundations Division*, proceedings of the ASCE, Paper 10009, vol. 99, SM9, September 1973, pp. 649-667.

49. Rosenberg, N. J., "The Influence and Implications of Windbreaks on Agriculture in Dry Regions," in *Ground Level Climatology*, R. H. Shaw, editor, Publication No. 86 of the American Association for the Advancement of Science, Washington, D. C., 1967.

50. Scott, R. F., "Heat Exchange of the Ground Surface," *Cold Regions Science and Engineering, II-A1, Part II: Physical Science, Section A: Geophysics*, U. S. Army Materiel Command, Cold Regions Research and Engineering Laboratory, Hanover, New Hampshire, July 1964 (49 pp.).

51. Sellers, W. D., *Physical Climatology*, University of Chicago Press, Chicago, 1965.

52. Shanks, R. E., "Altitudinal and Microclimatic Relationships of Soil Temperature Under Natural Vegetation," *Ecology*, vol. 37, no. 1, January 1956, pp. 1-7.

53. Shreve, Forrest, "Soil Temperature as Influenced by Altitude and Slope Exposure," *Ecology*, vol. 5, no. 2, April 1924, pp. 128-136.

54. Singer, I. A., and R. M. Brown, "The Annual Variation of Subsoil Temperature about a 600' circle," *Transactions* of the American Geophysical Union, vol. 37, no. 6, Dec. 1956, pp. 743-748.

55. Stanhill, G., "Observations on the Reduction of Soil Temperatures," *Agricultural Meteorology*, vol. 2, no. 3, pp. 197-203.

56. Stephenson, D. G., "Insulation to Prevent Ground Freezing," *Building Research Note* No. 119, Division of Building Research, National Research Council of Canada, Ottawa, April 1977.

57. Turnage, W. V., "Desert Subsoil Temperatures," *Soil Science*, vol. 47, pp. 195-199.

58. West, E. S., "A Study of the Annual Soil Temperature Wave," *Australian Journal of Scientific Research*, 5, 1952, pp. 303-314.

59. van Wijk, W. R., and W. J. Derksen, "Sinusoidal Temperature Variation in a Layered Soil," in *Physics of Plant Environment*, W. R. van Wijk, editor, North Holland Publishing Co., Amsterdam, 1966.

60. van Wijk and D. A. de Vries, "Periodic Temperature Variations in a Homogeneous Soil," in *Physics of Plant Environment*, W. R. van Wijk, editor, North-Holland Publishing Co., Amsterdam, 1966.

61. Williams, G. P., and L. W. Gold, "Ground Temperatures," *Canadian Building Digest* 180, Division of Building Research, National Research Council of Canada, Ottawa, July 1976 (4 pp.).

62. Williams, G. P. and L. W. Gold, "Snow Density and Climate," Engineering Inst. of Canada *Transactions*, vol. 2, no. 2, May 1958, pp. 91-94 (also available as publication NRCC 4833, National Research Council of Canada, Ottawa).

63. Willis, W. O., P. J. Wierenga, R. T. Vredenburg, "Fall Soil Water: Effect on Summer Soil Temperature," Soil Science Society of America *Journal*, vol. 41, no. 3, May/June 1977, pp. 615-617.

64. Yoshino, M. M., *Climate in a Small Area: An Introduction to Local Meteorology*, University of Tokyo Press, Tokyo, 1975.

APPENDIX A

Thermal Properties of Rock, Soils, and Concrete*

Material Description	Density lb/ft2	Specific Heat Btu/lb-F	Conductivity Btu/h-ft-F	Diffusivity ft2/day
Wet Soil†	117	.30	1.4	.96
Average Rock	175	.20	1.4	.96
Dense Concrete	150	.20	1.0	.79
Average Soil**				
Heavy Soil, damp	131	.23	.75	.60
Heavy Soil, dry	125	.20	.50	.48
Light Soil, damp	100	.25	.50	.48
Light Soil, dry	90	.20	.20	.26

* From ASHRAE *1978 Applications Handbook*

† From ASHRAE *1974 Applications Handbook*

** average thermal diffusivity, determined by Kusuda & Achenbach, 1965

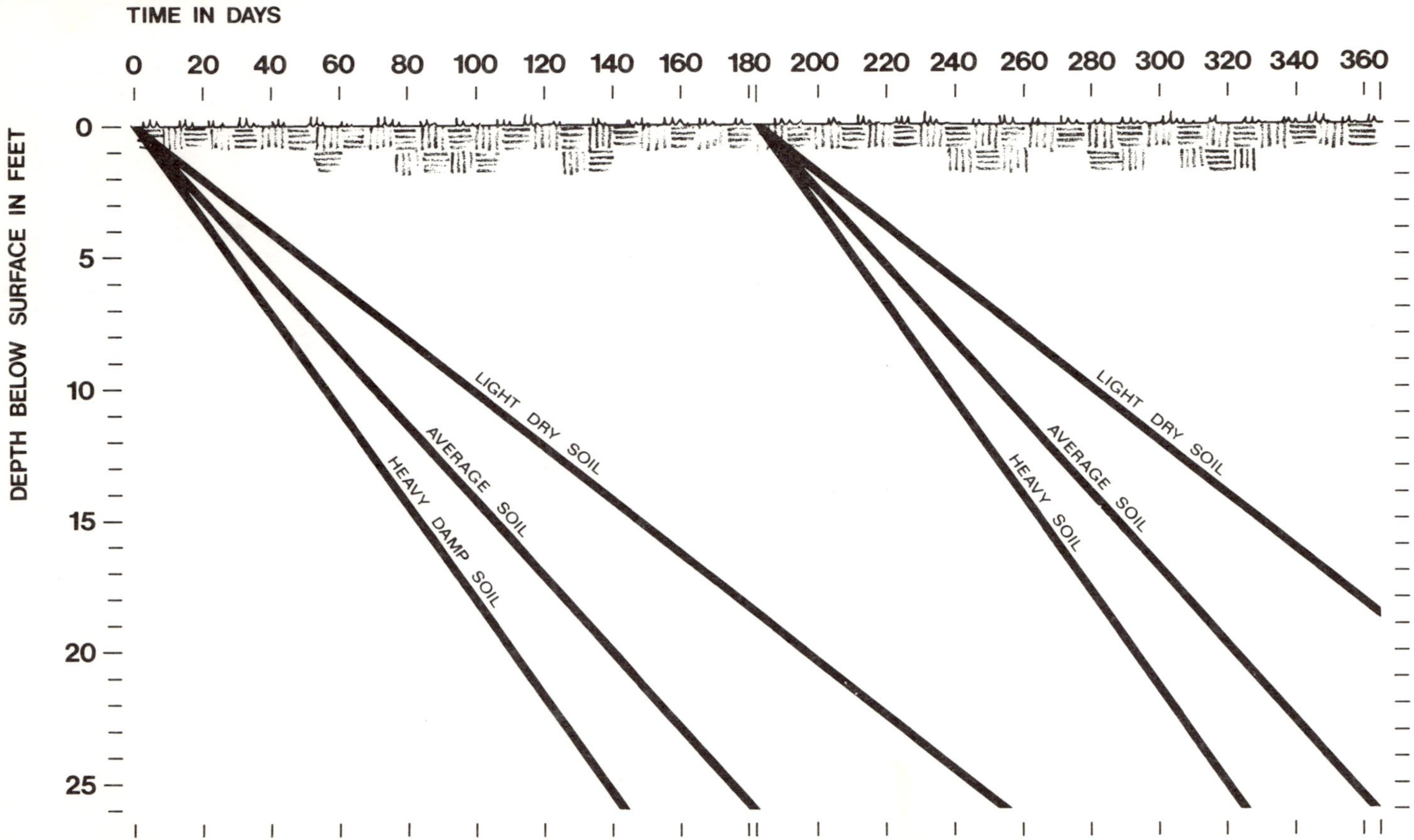

For a homogeneous soil, time lag is a linear function of depth and soil type. For a soil of average diffusivity ($K = 0.6$ ft^2/day), the lag is one week per foot depth, although this may vary slightly with seasonal changes in soil moisture content.

Troglodyte Communities in France

David Lord
Associate Professor
School of Architecture
University of Maryland

INTRODUCTION

The word troglodyte conjures up childhood images of gnomes, trolls and leprechauns to some people. The definition of a troglodyte comes from the Greek, *troglodytes*, "One who creeps into holes; cave dweller." The expression refers to any of the prehistoric people who lived in caves or to those who choose to live alone in seclusion. A less-used meaning refers to anyone who lives in a "primitive, low, or degenerate fashion."

Far from these definitions is the richness of life and culture of the underground dwellers of central France who live in the *maisons troglodytes*. The people who live here are the modern French who have inherited centuries-old homes which require very little energy or maintenance to keep them comfortable. To understand how a civilized quality of life and high standard of living can be enjoyed in pre-historic shelters, we might regard the thoughts of French President and sometime-philosopher Giscard D'Estaing who asked his countrymen to consider the silver lining as they tightened their belts in adjusting to economic inflation. He said that, "It is a good thing that the consumer society is doomed by international economic conditions, because it is not suited to the genius of France. The Sober Society towards which we are headed is more suited to French culture which has a sense of time and measure and to France, a country of rural values that preserves its attachment to the land."[1]

The adjustment to increased costs of imported goods including energy has been taking place quietly in the French countryside for many years as Frenchmen have re-discovered the homes of their ancestors.

Underground living and industry in France has existed from paleolithic and neolithic times continuously through recorded history. At Roc en Pail, fossil remains of prehistoric *homo sapiens* have been found in collapsed caves.[2] The first record found of the existence of troglodyte merchants of wine in the area studied is in the 14th century, drawn from the Saumur archives.[3]

Underground architecture has had its influence on modern architects. One example is the Paris U.N.E.S.C.O. Headquarters by Marcel Breuer (fig. 1) which has an administrative wing situated entirely below the level of the site, each office facing a central courtyard which delivers light and air to the interior spaces. The importance of underground locations at a time of national crisis is underscored by the subterranean bunkers and support facilities at the Maginot Line of World War I. (Fig. 2). A mechanized

Figure 1. Paris U.N.E.S.C.O. Headquarters by Marcel Breuer.

military city located 325 feet below ground was proposed but only partially completed. In the French commercial world it is interesting to note the reference to underground community in an advertisement for landscape gardening. (Fig. 3).

Troglodyte dwellings are distributed throughout the country. (Fig. 4). They occur in the highest density and largest numbers along the Loire River Valley, the region of the chateaux of France. It is precisely because of the construction of the chateaux of the Loire that the opportunity for underground living on a large scale came about (Fig. 5). As the stone cutters extracted massive construction blocks of sandstone, leaving modular cavities behind, their families and the families of local farmers, watermen, and artisians moved into the "left over" spaces. The opposite process occurred as well. When one of these families needed additional space in their apartment, they would cut a new room and sell the stone which they obtained in this way for construction.

GEOLOGY AND TOPOGRAPHY

The geology of the region is typical of the area of France known as the Paris Basin. In this area, sediments were laid down during the Cretacean period when the ocean covered the area as far north as Rennes, and later

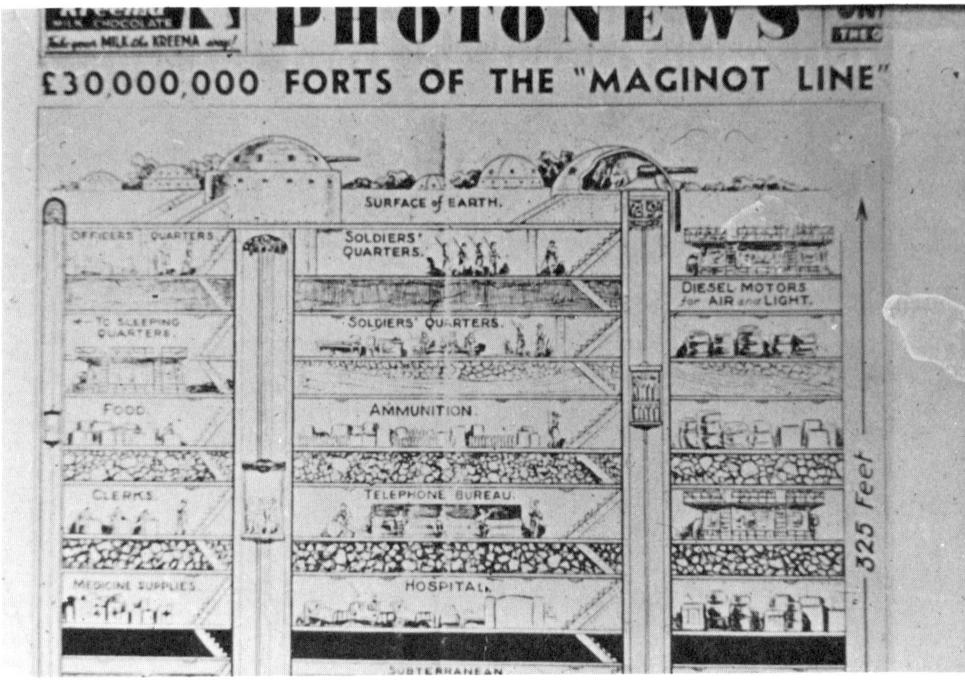

Figure 2. A report on the underground military structure at the Maginot Line.

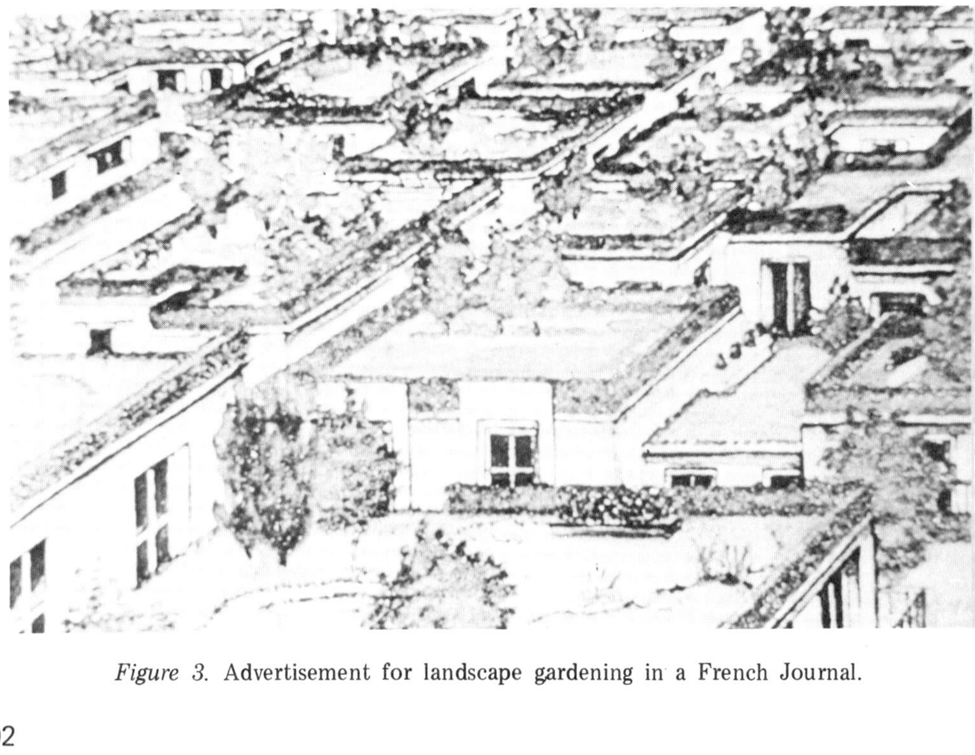

Figure 3. Advertisement for landscape gardening in a French Journal.

during the Miocene epoch, when the edge of the ocean was at Tour and Blois.[4] These sedimentary deposits, composed of shelly and chalky sandstone, overlain with shale and clay, were the beaches of ancient oceans but are now some 100 meters above sea level. The ceilings of the houses located here are often covered with the fossil record of their geologic origin. Such fossils as shells, pine cones and ammonites are found in abundance.

Figure 4. Map of France showing the major rivers and distribution of underground structures.

Figure 5. The Chateaux of the Loire Valley.

The slope of the sedimentary deposits is generally to the west. Erosion by water has led to the development of cuestas with exposed *tuffeau* cliffs which generally face the east. (Fig. 6). The *tuffeau* or chalky sandstone, is covered by a layer of fertile arable land which supports the agriculture of the region, consisting mainly of vineyards, orchards and livestock.

Doué-la-Fontaine, which is at the center of the region, is geographically located at 52 degrees north latitude at the longitude of the Greenwich Meridian. The annual rainfall for the area is 20 to 30 inches per year in a climate that is warm and humid in summer with no dry spell. Average January air temperature is 38F, average July temperature is 66F. The measured indoor temperature in subterranean spaces varies from 59F to 61F throughout the seasons.

The cuestas with their exposed cliffs, reveal stone which is 10 to 50 meters thick from which stonecutters have extracted building material over the ages. To the northwest, toward Anger, the quality of the stone is less desirable and tends toward greyness in color with less durability. In the area around Samur, the stone is whiter, cleaner, harder and more desirable as a building material, especially for exposed or finish construction such as cornices, facias, veneers, mantels and other architectural decoration.

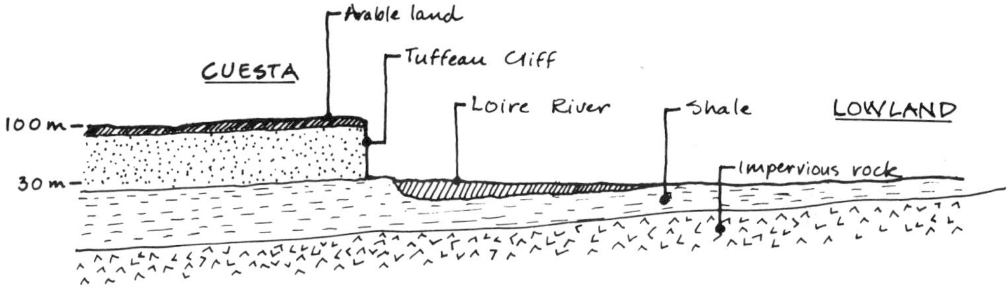

Figure 6. General geologic form of the Loire Valley.

If one draws a line on the map perpendicular to the Loire, passing through Anger, it defines the western limit of the geologic bed favoring the establishment of troglodyte colonies. The geographic limits are from north to south (fig. 7): from Tiercé to Villevêque, through Sarriqué, Seiches, Bauge, Cornillé, Andard, Beaufort, Alloué, along the left bank of the Loire, towards the east from Saint-Jean-des-Manurets to Montsoreau.

Underground dwellings occur in two major circumstances in this area: either on the face of the cliffs overlooking the rivers and streams or, more

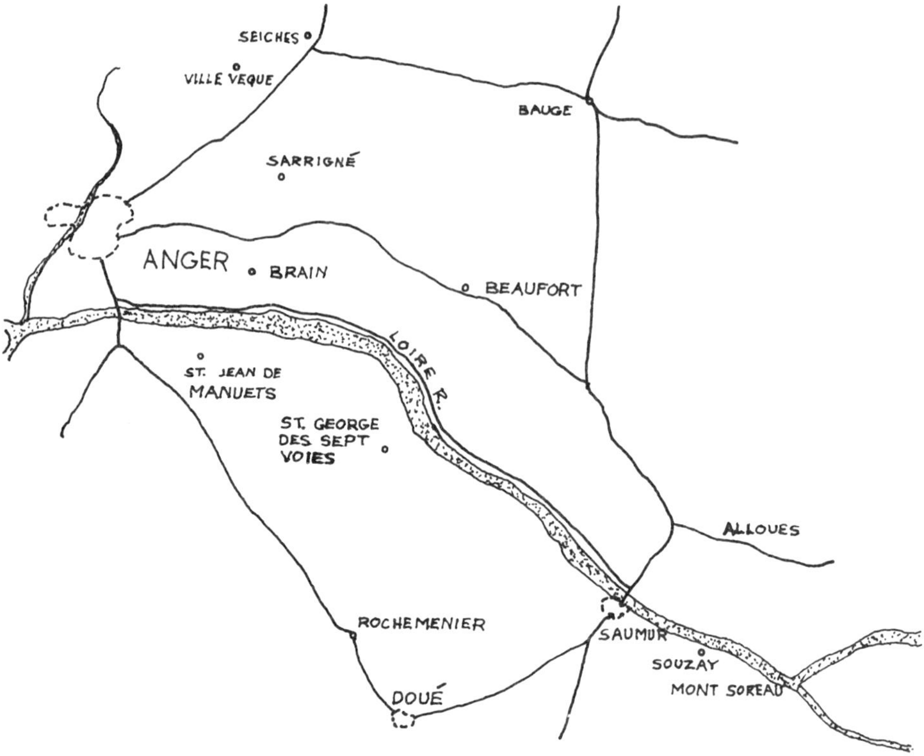

Figure 7. Map of the area where the highest concentration of troglodytes occurs.

surprisingly, on the surrounding backslope where there is perfectly flat terrain. In the latter case, an excavation forms a courtyard towards which the individual dwelling units and their outbuildings face. One example of the flat-terrain type is near the farm called *Les Touches*, near Montsabert. In the middle of a thicket there is an excavation of 15 meters diameter, 6-7 meters deep. A stairway cut from the rock descends to the house and its outbuildings. Windows and doors are cut from the stone. Air and light penetrate perpendicularly from openings in light wells. The form of excavation is different from those near Saumur. This is because the grey stone of the flat terrain is mined in closed quarries, not in open air.

About a third of the troglodyte population is in absolutely flat country, where access is vertical. In these dwellings in flat terrain, one sees openings facing all cardinal points, but wherever possible, the openings of windows and doors are toward the *Midi* or south of France.

The dwellings which overlook the river valley are the more dramatic and visible of the two types. These are more easily found by the casual traveller and they also seem to have more richness of architectural expression, at

Figure 8. Semi-underground structure found on the face of a cliff.

least on the facade. Here are also found semi-underground examples with partial construction occurring on the outside of the cliff (fig. 8). In plan, the two types are often similar in their meandering and free-form character.

The inhabitants of each of the two types of dwellings are of different occupations. The cliffs are the most active places in the quarrying and shipping of building stone. This is where the cutters and the watermen made their homes. The plains region is given over to farming and the structures here accommodate the activities of producing wine, harvesting nuts and fruits and sheltering grazing animals.

CULTURE AND DAILY LIFE

In the Loire valley, living and working underground has been a part of the culture of the region for such a long time that it has given rise to a folklore surrounding this style of life. The folklore has been passed on in the oral tradition through legends, stories and songs about the heroes, the bandits, the sorcerers, the hidden treasures, the romances, and unexplained crimes which have occurred over the years. The troglodyte caves have provided refuge to hundreds during time of civil unrest, during the religous wars and during the French Revolution.

Community life is reflected in the function of the architecture; there are building types that meet every need of the citizen. There are underground churches, schools, public buildings, and meeting halls (*velloirs*) where weekly dances were held. There exist the remains of a variety of commercial shops and artisian's workshops. Butchers, bakers, blacksmiths, wheelwrights, coopers, spinners and weavers were all lodged next to or within their shops or ateliers in the cliffs overlooking the Loire. The most numerous dwellers were the rockcutters and in each hamlet there were at least five or six of these with their families.

When they didn't live in the galleries where the rock was being mined, they would leave for their subterranean quarry, winter as well as summer, at about 5 a.m. The sound of their wooden shoes on the pavement served to wake the wives of the village. In the evening they returned at 8, working 11 to 12 hours at the end of the tunnel. If the stone was of good quality, they received five francs for the day.

Watermen were perhaps the next greatest in number and the houses of these people were often decorated with sculpture which had a marine theme (anchors, boats, fish) carved directly in the walls, usually above the mantel. There were the businesses which were attached to commerce on the waterways: the boatbuilders, rope makers, shipping agents, hoteliers, carriage builders, glass bottle and pottery makers for the shipping of wine and oil from the region.

Little industries which were sequestered in the caves were recorded by a visiting Englishwoman in 1906: "I visited after dinner a glassmaker's factory and everywhere there were glass toys. In a hall carved in the rock, three workmen were seated at a table..."[5] Potters made the containers in which oil was stored. On the fourth of January, 1630, at seven in the morning, the curé of the parish of Fontevrault gave the last rites to a potter practicing his craft in an underground home.[6] At Cornillé-les-Caves at the troglodyte hamlet of Perruchère, with 30 homes, one still finds the cave of the hairdresser. The craftsman or artisan, unlike the farmer, required little space. The house of the quarrier was one room only. The weaver had a supplementary cave. The same for the blacksmith, butcher, shoemaker, mason, etc.

Other artisians who were connected with the agriculture and economy of the region were the basketmakers—baskets were and still are made from cane, reeds and grasses of the region—spinners (of wool), weavers, bird trappers, and artisians who carved decorative and functional elements from the stone such as sarcophagi, lamps, mantels, sculptures, etc. There were those who made a living making lime by reducing limestone in ovens or kilns. Others mined potassium nitrate (saltpeter) which was used in the making of gunpowder. A significant building type in the middle-ages and renaissance was the dovecote (*fuie*) where tens of thousands of pigeons roosted at night, leaving valuable deposits of fertilizer and providing meat for the table of the elite, since it was only the nobility and landowners who received permits to build these structures.

One of the interesting community spaces was the *veilloir*, or evening gathering place. this was a space of at least 20 meters in diameter, around which a bench 50cm. above the floor was carved out of the rock. More than 100 couples could dance all night to the sound of a violin and accordion. At Denezé-sous-Doué (5km. north of Doué) sheep stables were used for this purpose. The beasts were pushed to another stall and people occupied the space which had been pre-warmed by the flock. Lighting came from a single source, originally a large candle of resin, costing one *sou*. Later, a candle of tallow (two *sous*) was used. For thermal comfort, many of the *veilloirs* were stocked with layers of straw which was gathered up around the legs upon being seated.

ARCHITECTURAL FEATURES

The manner of extraction of the stone left spaces which were suitable very nearly as they were to serve as the interior of dwellings occupied by stonecutters, watermen and artisans. All that remained to be done after quarrying was the filling-in of the facade with masonry. In fig. 9, the shape of the quarried stone, [(1)a], and the left-over space, [(1)b], are shown.[7] This space is tailored to individual needs by adding a floor two meters above the existing earth floor. This provided a space for storage and for

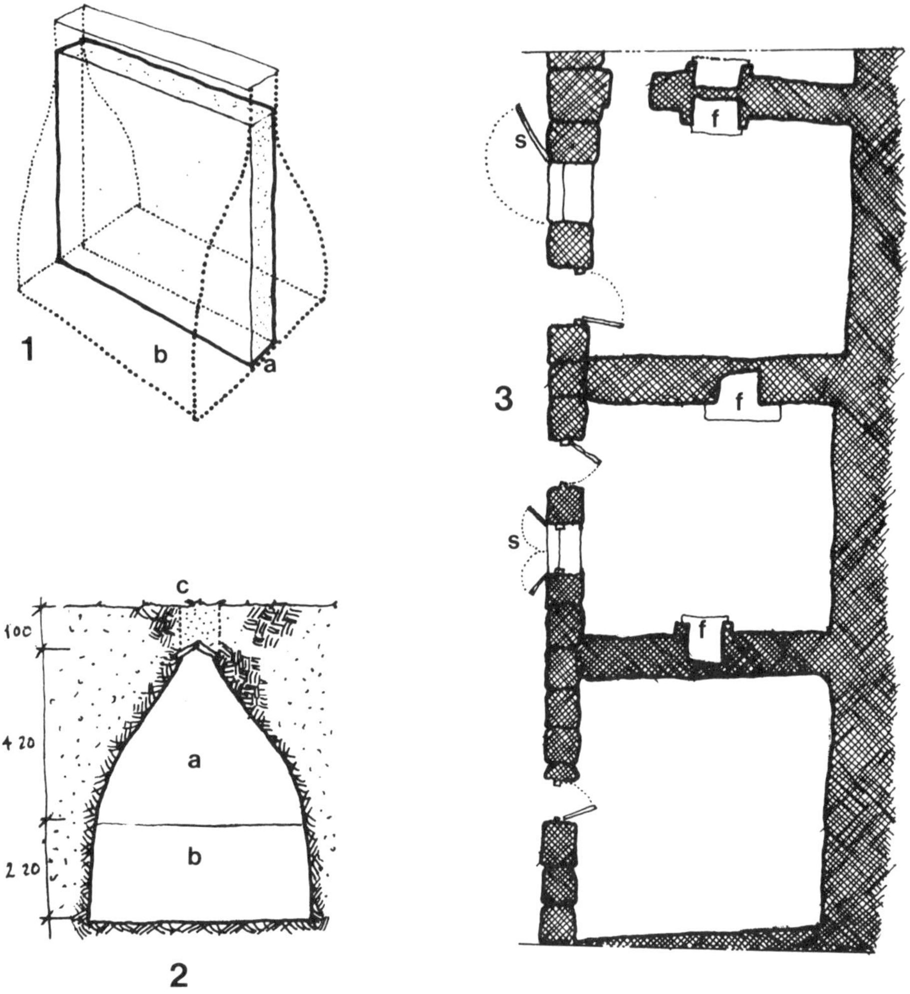

Figure 9. The extraction of stone gives a nearly-filled space which can be used as a dwelling.

keeping animals. The living space, [(2)a], is a little over four meters high. The opening at the top of the space is closed off in the winter by a roof of tiles covered with straw, [(2)c]. The plan of an aggregation of such spaces is shown at the right of fig 9. Shutters, [s], and fireplaces, [f], are common to the type of structure. A sketch of the interior of such a house is shown in fig. 10.

A more developed apartment carved out of the cliff is shown is fig. 11. The plan shows a narrow, elongated house with a fireplace and two floors with a connecting stairway. The first of the levels is used for storage or for animals. A planked floor, [3], and the matrix of sandstone separates the occupied floors. The facade of a one-story series of dwellings is shown in figure 12.

Figure 10. Interior of an apartment in the left-over quarried space.

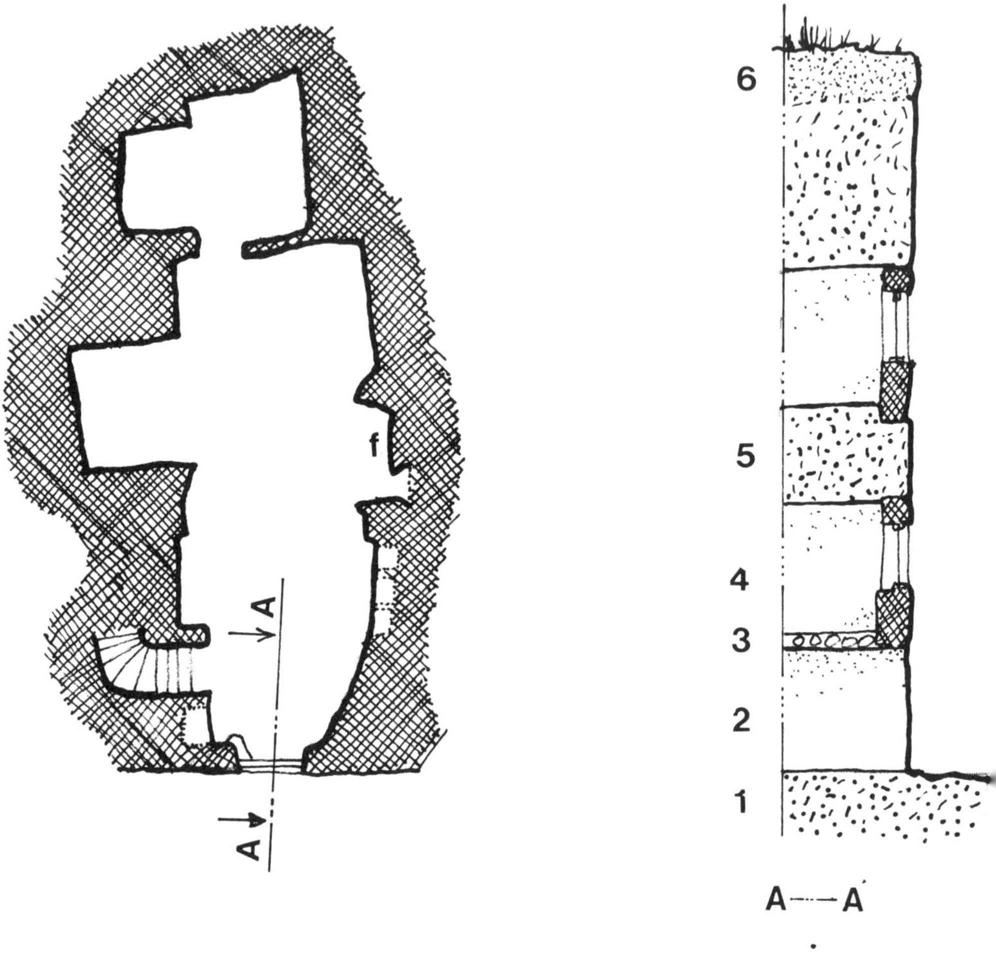

Figure 11. Plan and section of an apartment carved out of a cliff.

The shelters occupied by the farmer were considerably more complex in nature. Each underground farmhouse usually had its own source of water, a well, located either in the courtyard or sometimes inside the home. In addition, there were one or more ovens and fireplaces for baking and cooking and as a source of heat. Many places had a special room for the *pressoir* for reducing the grapes at the time of harvest. Sometimes there was also a *huilier* for the pressing and extraction of oil from nuts. These *huiliers* were used clandestinely during World War II when local residents lived off the land during times of rationing of food.

Figure 12. Facade of a one-story series of dwellings carved out of a cliff.

The plan in fig. 13 is of a typical farm dwelling which occurs in perfectly flat terrain near Doué-la-Fontaine. All the spaces face a courtyard [A]. The dwelling spaces are at [B]; stables and storage at [C], and [D]; the *pressoir* is at [E]; the ovens and fireplaces at [F].

A larger and more contorted structure is the underground church at Doué-la-Fontaine, which was known to exist in the last century and probably existed earlier than that (fig. 14). The church is essentially a long cave extending north-south, about 80 meters long, composed of a large number of chambers, connected by a vast collection of underground corridors, some of which are shown in plan. The corridors extend

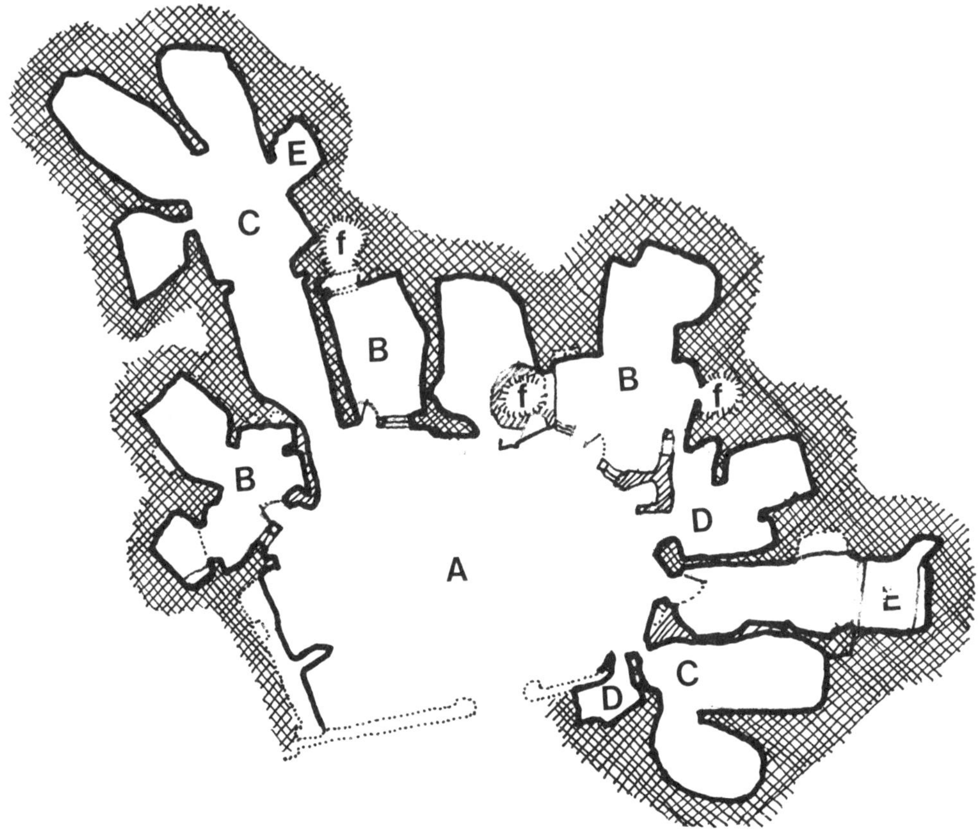

Figure 13. Farm dwellings and outbuildings which occur on flat tarrain.

horizontally as well as vertically, allowing access to other apartments or refuges. Over the course of time, many corridors have been walled up, so only part of the entire cave is shown. In fig. 14, the entrance [a] is located in the courtyard if a building which now houses a hardware store (Lombard). Continuing along the entry corridor, one descends and passes under the entry along an inclined passageway to a door under a semicircular arch, opening to room [B] provided with a well, "w_1" of square cross-section. The vault of this room is supported by an arch pierced at the corner by a circular cavity "v" (1.8 meter diameter) probably a ventilation shaft. Another excavation at the top of the vault situated at the left of the entry door, leads to other galleries circulating above the church.

Adjacent to room [B] is room [A], oriented NNE-SSW and which has a vaulted ceiling (2 meters, 80 high) carved like a basket handle and containing another well, "w_2". These two rooms probably served as a

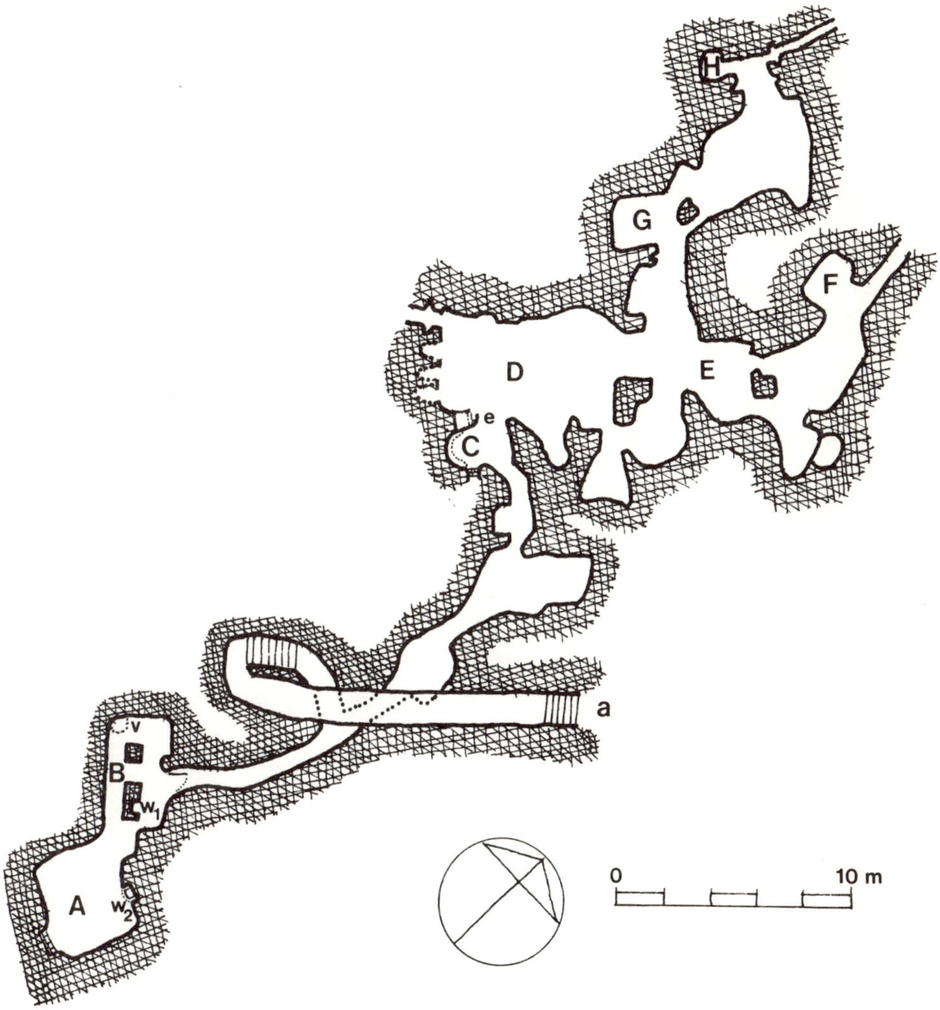

Figure 14. Plan of underground church at Doue-la-Fontaine.

kitchen at one time. A little tunnel 1 meter 20 high by 55 cm. wide, which is walled up today, leads towards other caves. Proceeding towards the north, along the curridor, we come to a little room [C], shaped like a cupola with a semicircular bench carved out of stone, 1 meter 50 above the floor and 50 cm. wide. It is provided with a bay opening into the church.

The opening "e", 1 meter 90 wide with a half arch is very recent and opens into the church central space, [D]. Here we are about 10 meters under the surface of the land. This hall is oriented SW-NE and is 14 meters long. The vault is 5 meters high. The features which add to the religious

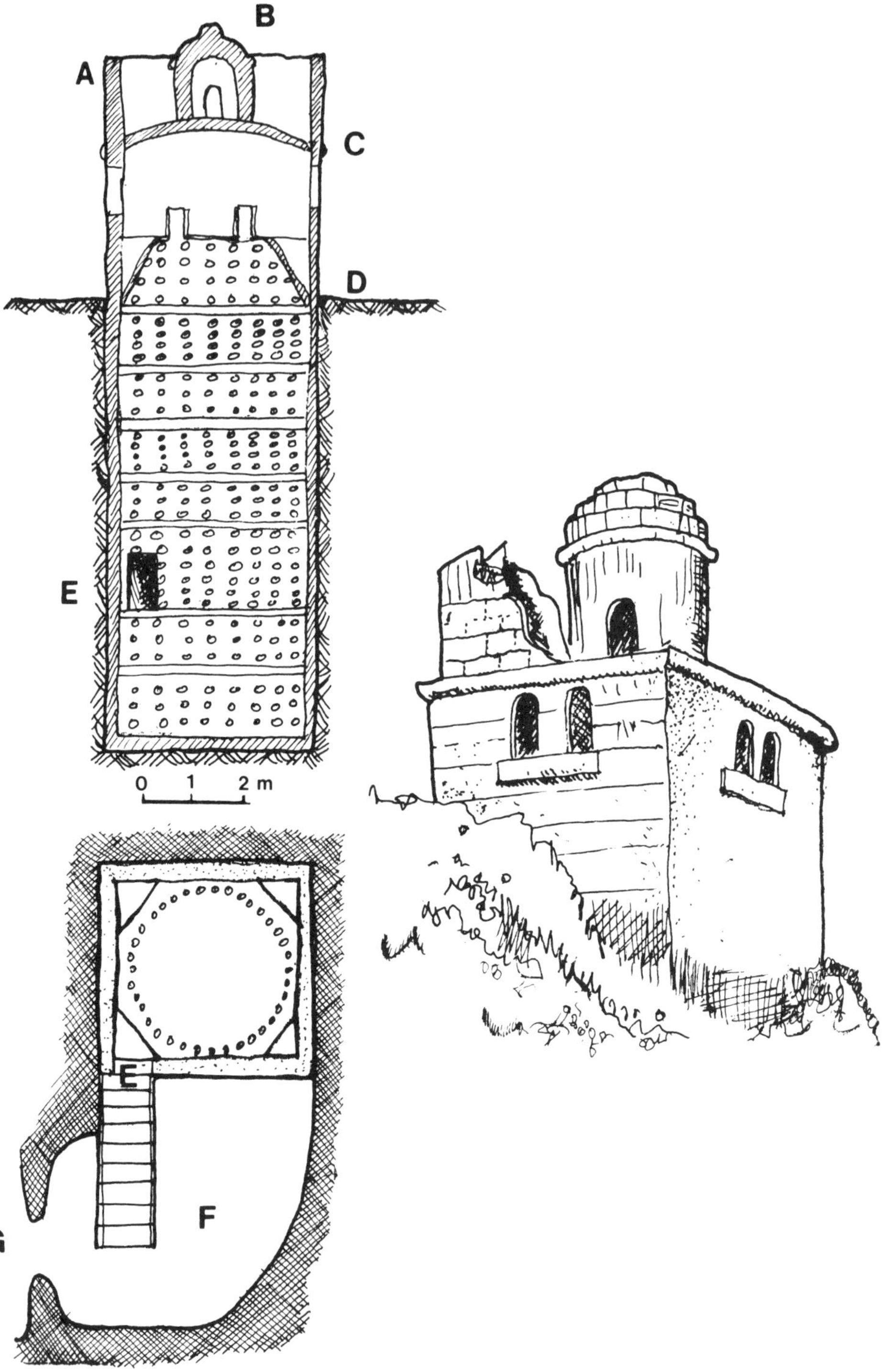

Figure 15. Dovecote near Vieux Logis.

character of the space are the windows at "d" beginning at 1 meter 50 above the floor and extending to 5 meters high. They are 4 meters 80 wide. These openings have been closed at the top but formerly extended to the sky above. In the center of the panelling is a little excavation which served as the tabernacle. At "n" a niche holds a lead bowl used as the holy water font. A tunnel leaves the room at the SW corner and a well, "w_3", 2 meters in diameter, completes the description. At the NW, there is an arch which forms a monumental door to the church. Continuing in this direction there are a succession of spaces with their own wells.

Another interesting structure is the dovecote shown in fig. 15, located near Vieux Logis. This structure is very nearly in ruins, but is used by pigeons to this very day. The tower extends 5 meters above ground level [D] and some 8 meters below ground. The returning pigeons enter the tower and descend to their roosts. Human access is through an underground portal at [G] through a doorway, [E], and along ledges inside the tower which allow one to climb to various levels.

This brief survey of some underground spaces in France has only touched the surface of a rich heritage of indigenous architecture in which people have lived, worked and played underground and in harmony with their environment for centuries. Perhaps the re-discovery of this culture will help shed light on our own needs and aspirations as we search for new directions.

REFERENCES

(1) Koven, Ronald, "Letter from France," *Washington Post*, September 21, 1979, p. A22.

(2) Gruet, Docteur, "Restes d'un homme fossile en Maine-et-Loire," *Bulletin de la Societe d'Etudes Scientifiques d'Angers* Tome I, 1950, p. 19.

(3) A proclamation dated 1388 by the judge of Saumur confirming that the underground abbey of Saint Maur had the right to sell wine.

(4) Debelmas, J. *Géologie de la France*, 2 vol., Doin éditeurs, Paris, 1974, Vol 1, p. 232, Fig 100.

(5) Craddock, Ms., "Voyage en France," *Jn. Anjou Historique*, Jan./Feb. 1907, p. 372.

(6) Civil archives of Fontevrault, series GG, G7.

(7) Guyomard, G., "Les Troglodytes en Saumurois," *Techniques et Architecture*, no. 325, June/July, 1979, p. 96.

General References:

Strahler, A. N. and A. H. Strahler, *Environmental Geoscience: Interactio between Natural Systems and Man*, Hamilton Publishing Co., Santa Barbara CA, 1979.

Davies, G. M., *A French-English Vocabulary in Geology and Physica Geography*, John Mann, Publishers, London, 1932.

Fraysse, J and C. Fraysse, *Les Troglodytes en Anjou* à *Travers les Ages*, 3 vol., 2nd. Ed., Farré et fils, Cholet, France, 1963.

"Une Maison Troglodyte," *Maison et Jardin*, no. 37, 1956.

ISOlation, INSUlation, INSOlation – Earth Contact as Heat Transfer

Gordon L. Moore
Professor of Mechanical and Aerospace Engineering
University of Missouri – Columbia
Columbia, Missouri 65201

The order of listing of the three principles in the above title is the theme of this paper. I will show that, if the isolation principle is followed properly, the need for insulation is substantially reduced, and the need for insolation is so greatly reduced that only a token amount of direct solar input is needed.

On the other hand, it can be easily shown that, if these three principles are accented and followed in the reverse order, the resulting design will fail, will not result in energy saving, and due to solar overkill, may wind up as a solar oven or as a people cooker under certain winter conditions. We have two examples around Columbia and Kansas City.

EARTH CONTACT IS A FUNCTIONAL, ACTIVE PRINCIPLE

Others who are working in the field will give various reasons for using such terms as "subterranean", "underground", or "earth-sheltered." Let's bypass all the semantics and look at what we want the home to *do*. The bottom line is that the genus we are discussing here has to *contact* the earth with a major percentage of the building's exterior surface area if it is going to work, i.e., gain independence from outside fuel sources while maintaining human comfort.

There are many examples around our area which have followed this principle of utilizing 75% or more earth contact. All of these have achieved 75% to 90% independence of conventional fuel inputs. Unfortunately, there are a few others which have used only one earth berm, or only 2 or 3 feet of berm on the 8 foot walls. All of these have been, and are, energy failures, very little different from suburbia's all too prevalent energy sieves (houses). For these and other reasons, I will use the term "earth contact" not as a generic, but as a descriptive, functional adjective.

THE ISOLATION PRINCIPLE HAS A COROLLARY

From the heat transfer viewpoint, you can look at the earth contacting structure as a heat source and at the surrounding earth as a heat accumulator and delayer. When you see the profile of such a building inset into the earth, you can easily recognize which parts of the building's exterior surface will most readily give off heat *to* the surrounding earth mass and also *through* the intervening mass to the distant surface. Note the use of the term "earth mass".

If you think of the surrounding earth as a means of *isolating* the warm living space from the temperature extremes at the distant earth surface, it seems reasonable that the length of the heat flow path *from* the living space *to* the earth's surface would have an important effect. It does. Let's look at it.

The path length from the warm structure to the cold ground surface not only moderates the temperature extremes travelling from the surface, but also delays the travel of temperature changes at a rate of about one week of delay per foot of path length. For example, a floor area near an 8 foot, earth covered wall would have a curved heat flow path leading first downward and then outward and upward to the ground surface. This particular path might have a total length of 12 to 15 feet. Thus, a very cold temperature extreme would take 3 or 4 *months* to travel down and affect the floor temperature, and then only with a negligible amplitude.

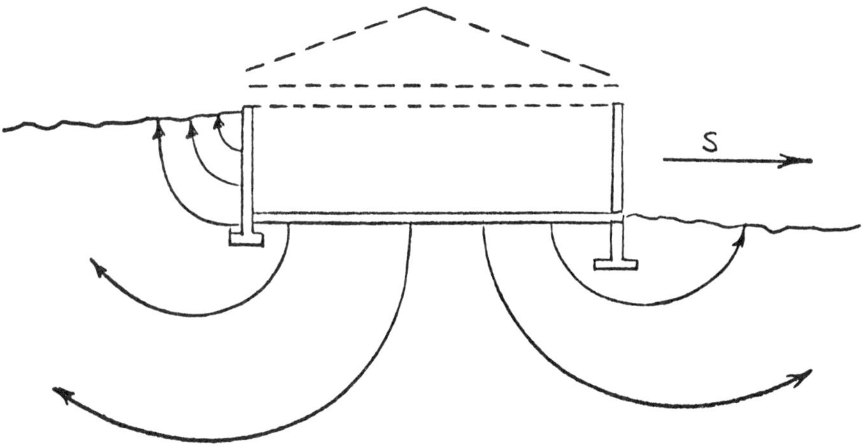

Figure 1. Earth Contact House Cross Section Illustrating Heat Paths.

THE LONG PATH PRINCIPLE APPLIED

I have coined the term "Long Path Principle" to describe this time delay effect. Note in Figure 1 how the heat flow lines are shaped around the earth contact structure, and also observe the relative lengths of the heat flow paths from the various parts of the structure to the ground surface.

By the simple expedient of selecting external insulation placement for the purpose of lengthing the heat flow paths that were initially too short, anyone can arrive at a design with fewer cold areas in the walls and floors.

As an example, if we apply the long path principle to the structure sketched in Figure 1, it is immediately obvious that the best insulation placement would be along the outside of the upper portion of the north earth covered wall and along the exterior of the foundation wall at the south perimeter. This potential flow path from the south part of the floor can be lengthened further by placing a 2 or 4 foot width of rigid foam insulation under the floor, along the south, exposed edge.

Since most of the remaining interior floor areas have relatively long flow paths, there is no need to insulate under the entire floor. Our temperature monitoring programs around Columbia have verified this. One such structure, with no insulation under most of its area, has had floor temperatures holding at 65 to 68F during the two recent coldest winters on record.

THE HEAT STORAGE-HEAT RECYCLING EFFECT

A very significant effect either neglected or unknown to many earth contact designers is the slow but steady accumulation of heat along the flow path from structure to the surface. During the first few months of building occupancy, the soil mass nearest the structure will rise toward an equilibrium temperature gradient. If the interior space is kept at about 70F, the interior wall surface will approach an average of about 68F and the exterior surface about 65 to 67F. The soil temperatures will show a gradually diminishing gradient in the outward direction. Our observations and temperature monitoring of local earth contact structures has shown that this heat accumulation by the surrounding earth is not a steady, continuous process. There are some very pleasant surprises in store for those who use earth contact or observe this effect for the first time.

We found that this storage effect will continue but gradually diminish during summer. At about August-September, the flow will stop and reverse direction. The particular time will depend on the summer temperatures maintained in the space. At this saturation or changeover point, the surface temperatures will hold steady for a long time, and then slowly start to drop. With the onset of fall and the slow cooldown of summer heat, the earth contact walls will begin to gradually release their warmth back into the living space. This effect is so gradual that it will continue on into December. Here in Missouri, at least, the earth contact home heating season starts very gradually in mid-December and tapers down in mid-March, a three month heating season! Eat your heart out, suburbia, in your splendid energy sieves.

Stated in heat transfer terms, the earth mass around the structure serves not only as an isolater, but also as a heat recycler, feeding gentle warmth back into the living space during the fall.

Several of us in the Missouri Solar Energy Associates have long been curious about the energy flows in all the different types of earth sheltering. Ron Wantoch (solar consultant, Overland Park, Kansas) and I were each lucky enough to have a home computer, and decided to make computer studies of the soil temperatures around a typical, earth contact structure. Using moderate programming limits (8K bytes), we arrived at some reasonable looking two dimensional plots of temperature lines (isotherms) around and below the earth contact walls and floor. This study showed us *how* the heat was flowing, but not *how much* heat might flow from a given structure. However, by comparing these computer plots with actual soil temperature data measured at our soils temperature station on the University of Missouri campus, we felt that we were close to the truth.

Perhaps the most significant conclusion stemming from this preliminary computer study was that the earth contact structure, when kept at interior comfort temperatures, will have a dominating effect on nearby soil temperatures. In order to accurately compute heat flows from an earth sheltered structure, you have to know the diffusivity of the particular soil, or know the actual, nearby soil temperatures. Knowing the theoretical, undisturbed soil temperatures at a distance from the structure is of little use. Average ground water temperatures are of even less use.

USING EXISTING SOIL HEAT FLOW INFORMATION

At this point we recognized that reasonably accurate earth contact temperatures could only stem from experimental measurements encompassing at least one year of time and extending downward at least 8 to 10 feet into the earth. Fortunately, such experimental data exists in at least one journal.[1]

These year-long measurements of temperatures and flows around an earth covered structure were made by ASHVE under weather conditions similar to those typical to Missouri. Our computer plots gave a reasonable match with the empirical measurements by ASHVE, so we were encouraged enough to assume that the heat flow rates measured by ASHVE could serve as estimators for heat flows in other structures undergoing similar environments. To verify this, we tried this method on an earth contact home near Columbia, Missouri.

Using the dimensions and properties of this 2400 sq. ft. earth bermed home, we calculated the losses through the non-earth contacting wall and ceiling using standard methods and R-values. For the earth contacting areas, we used the heat flow rates measured in the ASHVAE tests. The January total rates estimated were about 150,000 Btu/day, or about 180,000 Btu/day including infiltration losses.

This same house has had the same fuel use for the past two winters, i.e., about one cord of wood per winter. This is about 28,000,000 Btu of available energy. Using a 60% efficient stove, the net heat output from this stove would be around 15,000,000 Btu for the 90 day heating season, or about 170,000 Btu per day average. That is "in the ball park."

Some further verifications of our findings are evident in other papers in this volume. While we in M.S.E.A. and in Missouri are continuing the study of earth contact buildings and how they perform, I would like to see other earth contact builders and occupants do some additional monitoring of interior surface temperatures and fuel use. We need more empirical data. This information will be of great use in encouraging others to follow the earth contact route for energy saving and energy security.

ENGINEERING GUIDES AVAILABLE

I have initiated a series of Engineering Guides to inform the beginners on the nitty, earthy, gritty how-to-do-it-right in earth contact design and building. The two listed below may be obtained by writing the Industrial and Technical Referral Center, 1020 Engineering, University of Missouri, Columbia, MO 65211. Others are in process and will be available later. Missouri residents may pick these up at their County Extension Centers.

No. 5520: "A groundswell that is Sweeping the Midwest—Facts and Fancies"

No. 5521 "Water and Vapor Control in Earth Contact Homes"

REFERENCE

[1] Houten, F. C., *ASHVE Transactions*, No. 1213, 1942

Structural Design for Poured Concrete Earth Contact Structures

Willard S. Norton
Norton & Schmidt
Consulting Engineers, Inc.
1009 Baltimore – 8th Floor
Kansas City, Missouri

All of the materials used in above-ground construction can be used in building earth contact structures. However, the greater stress that these structures endure cause some materials, such as wood and concrete block, to be less practical than concrete. A thorough study of the building site, building owner's requirements, and the availability of materials all play a part in selecting which construction method to use.

Poured concrete for earth contact building has the edge over other less dense materials because of its superior strength and because it is naturally waterproof. Because of my continued excellent experiences with poured reinforced concrete earth contact structures, I will primarily discuss the tested, satisfactory use of this material.

Since the concrete used in earth contact structures is not the primary waterproofing material but is always covered with additional coatings, a medium strength concrete mix without additives, containing 5½ sacks of cement per yard, is quite satisfactory. Since it is not practical to use reinforcing bars smaller than No. 5 for stress control, it is recommended that these and larger bars be of ASTM 615 grade 60 steel while No. 4 or smaller bars be of grade 40. This lower grade for the smaller bars allows easier bending for stirrups and temperature steel. Temperature steel is important in controlling shrinkage cracks and should be installed to provide a minimum of 0.002 of the cross sectional area of slabs and walls. The temperature steel also assists in distributing stress resulting from differential settlement. The higher strength steel for the main reinforcing allows fewer bars to be used, thereby lowering the cost for this part of the structure. Considering the advisability of avoiding long slab spans that could result in excessive long term deflection, and the desire to retain maximum thermal mass, the 5½ sack mix is proving to meet the requirements of the industry. As with any material that uses water as a vehicle for hydration, it is very important that the water-cement ratio be maintained in accordance with accepted industry standards. Abuse of the recommended water-cement ratio will shorten the useful life of the structure and in extreme cases may result in structural failure. The introduction of fly ash as an additive to the concrete mix adds to the indeterminancy of the concrete and should not be used. Since concrete suppliers are using fly ash as a matter of course when not specifically asked not to, special attention regarding this matter is required.

The one difference between the use of concrete as just described and post-tensioned use is the replacement of mild steel reinforcing with post-tensioning cables. The temperature steel requirements remain a factor that calls for the use of mild steel. Post-tensioning, if used under the right circumstances, can be cost effective. These circumstances include the availability of trained personnel and the demand for a sufficient quantity to make its use practical. One ever present problem in the use of post-tensioning is the possibility of breaking a strand. These items introduce complications into what should be left uncomplicated. Also, few builders of small structures are equipped to post-tension concrete. However, as earth contact structures begin to play a larger role in providing habitable space, post-tensioning for slab construction will become a favorable way to build.

Given an available supply, precast roof slabs for earth contact buildings offer a practical method of construction. The practicability lies in that the precast units serve the structural requirements and provide formwork for the placement of a minimum 4 inch concrete top slab. This top slab is recommended to provide an even base for the installation of the waterproofing material. Also, the poured top slab provides for weight distribution. This slab should be reinforced with a welded wire mesh. The use of precast units for roof slabs is especially suitable for individuals who want to do their own construction. Contractors that specialize in basements are available in large numbers, so the do-it-yourselfer can easily subcontract the components that will give him a shell that can be covered ready for interior finish. If the individual waited in line for the total poured concrete structure, it could be a year or more before the shell is finished—not to mention the high profit that builders with backlogs can command. At the present time, some total poured concrete shell contractors are able to command an overhead and profit amount of 35%. Saving this by serving as your own general contractor can go a long way toward reducing costs and finishing the structure in a reasonable time. It is not advisable, however, for an untrained person to assume the responsibilities of being his own general contractor without the guidance of an experienced consultant.

Aside from the roof structure, the soil that the earth contact shell is bearing on is the most important aspect of an earth contact home. If the soil bearing strength is not adequate, the shell will be extremely costly to maintain and in severe cases will not be habitable. Failures such as broken walls, buckled floor slabs and water leaks will result from overestimating soil bearing capacity. Since it is presently considered acceptable to build above ground dwellings without the aid of soil tests, it is natural for the same attitude to prevail with below ground structures. There is a vast difference in that the weight of concrete underground structures is so much greater than conventional above ground houses that soil bearing requirements are equivalent to medium sized commercial buildings. It is seldom that an investor will risk his money on a commercial structure without a soils examination. A prudent builder of below ground structures would be well advised to employ a soils engineer. The average cost for an adequate soils report is a very small percentage of the value of the project. The soils report will give information about the soil bearing capacity and a recommendation about the most suitable foundation system. Often the report will give optional methods of supporting the foundation. The cost of the test is small compared with the costs that can be incurred if foundation problems develop. As in all things there are exceptions. If the foundation is going to bear on shale or rock then a visual examination by a competent indivudal will suffice, and the cost of soils analysis can be saved. The worst possible bearing condition would be on different types of material such as clay and rock or on soil with a high organic content. In both cases the foundation will not reach a static state without leaving the structure deformed and unacceptably damaged. A soils analysis will provide information on how to avoid damage from plastic soil. This type of soil expands when not utilized in a correct way. This expansion can do as much damage, or more, than that which is caused by severe settlement.

The surest way to avoid damage from settlement is to support the foundation on uniform material. If uniform material is used, and the footings have been designed for balanced bearing by an engineer, the settlement will be the same across the totality of the structure. All structures will settle to some extent, but if the finish items of construction are done correctly, the total settlement will not be obvious to the eye. It is essential that the concrete floor slab be free floating from any part of the concrete shell. If the slab is not free, the center of the slab will remain where it was placed while the sides and other parts that are attached to the shell will sink. This causes the buckling of floors and other finish items. Naturally, if the shell is supported by unyielding material, such as rock or shale, then this free floating condition is not a requirement. But, if the material is shale, provisions for preventing expansion will be needed. These provisions generally require good drainage to prevent water contact with the shale.

Since some movement between the shell and the floor slab will take place when the foundation is supported by compressible material, it is common sense to provide for this movement where finish items bridge between these two components of the building. One example would be wood partitions. If these partitions are placed snugly against the floor and the concrete roof above, settlement of the shell will crush the wood partition as the shell begins to distribute parts of its load onto these partitions. This can be avoided by leaving a ½ inch space either at the top or bottom of the partition. Understanding this requirement will provide clues for the handling of other items that can be damaged by settlement.

Not a finish item, but very important, is the construction of the garage approach slab and front door stoop. These two items of construction should be sloped to accommodate the anticipated settlement without the problem of a reverse slope toward the building. Attention to this can prevent the costly act of tearing out and replacing these approach areas.

An above ground dwelling can have a leaky basement and it may not be more than a minor annoyance. If the below ground dwelling leaks, it is intolerable. Extra care should be used in preventing water leaks. The very nature of the structure provides a good forum for preventing leaks because the heavily reinforced walls will not have cracks for water entrance, provided a good foundation bearing is present. Nevertheless, it is necessary to add to the leak-proof insurance by installing 4 inch drain tiles around the perimeter of the footings and, in some cases, under the interior concrete floor slab. This will provide easy runoff for any water that might collect against the exterior walls or under the floor slab. If interior drains are used, these drains should not be connected either to the sewer system or to the exterior drains. A blockage of the sewer or the exterior drains would run water directly into the area under the floor slab causing expansion of the soil under the slab with runious results.

Good drainage also serves the purpose of keeping lateral stress from building against the exterior wall. Since these walls are generally designed to handle an equivalent fluid pressure of about 40 pounds per square foot of surface, a doubling of this pressure from poor drainage could break the concrete wall. Also, poor drainage will cause a change in the moisture content under the interior floor slab, resulting in heaving. Often, this heaving is confused with foundation settlement. It takes a practiced eye to distinguish the difference. If any of these conditions occur it would be profitable to consult with an accomplished foundation engineer before any repairs are commenced. Although it is rarely done it would be an excellent move to provide for the cleaning of the foundation drain tile. These tile tend to silt up after fifteen to twenty years and, at that time, the unpleasant things described above may begin to happen. Providing for this cleaning would not involve much more than a strategically placed manhole or using curved corners so that a roto-rooter can be run the whole length of the tile without getting hung up on corners.

As described previously, the water-cement ratio is an important factor in developing strong, durable concrete but equally important is the curing time. The longer that the hydration can be extended, the more impervious to moisture the concrete becomes. This is because rapid curing leaves shrinkage cracks and surface shearing. These two items allow water to enter the surface resulting in the eventual deterioriation of the reinforcing just below the surface of the concrete. After the reinforcing begins to deteriorate, the rust separates the concrete on the face from the interior core of concrete leaving the entire section susceptible to even more rapid deterioration.

Although the underground structure is protected from this particular sequence of deterioriation, it is still well to know what improper curing does to the finished concrete product since anything that adversely affects the strength of this basic material is going to result in a shorter trouble-free life. The minimum curing period in order to accomplish satisfactory results is 4 days for walls and 6 days for slabs. This means that the forms should be left in place for at least that long. The walls can be stripped of their forms earlier than that if they are covered with burlap and kept continuously wet, or if the entire surfaces of the walls are sprayed with a curing compound. Since it is virtually impossible to find a situation where curing compounds are applied properly or where walls are kept continuously wet, it is recommended that the forms remain in place if possible. "If possible" is a big factor because concrete contractors need to have their forms off the walls as soon as possible in order to move to the next job and continue their livlihood. It becomes apparent that this reality of the industry is going to be a problem increasing in intensity as the pressure on the existing contractors increases. The equalizing factor can be in educating the user so that he picks up the responsibility of proper curing where the contractor cannot or is unwilling to perform. In any case, the forms for the roof slab must remain in place for at least 6 days. This is necessary because removing the forms earlier than that causes the reinforcing steel to work prior to the development of a minimal concrete bond around the reinforcing. This is known as "loss of bond" or slippage between the reinforcing steel and the surrounding concrete.

An early loss of bond sets the stage for early "creep" or deflection in excess of the design allowance. Avoiding this deflection is vital because of the heavy sustained loads of the earth cover and the slab itself. In cases of severe abuse in this regard, the roof slab would be judged dangerous and extensive repairs would be required to provide safe conditions. Because the beams are carrying even heavier loads, they should not have their forms removed in less than 10 days for the same reasons as described for slabs. Once the slabs and beams are stripped of their forms, they should be resupported at their mid-points with 4X4 wood posts cut slightly long and driven into place. These posts should be left in place until the fifteenth day. This extra time after the forms are stripped allows the concrete to develop the additional strength required to firmly hold to the reinforcing after its own weight is imposed. Attention to this particular aspect of concrete roof construction means the difference between a safe roof and one that is dangerous.

It is recommended that the reinforcing steel in poured concrete have minimum cover. It is important that the specified cover be adhered to. In cases of too much cover, the strength of the wall, slab, or beam will be affected and is situations where the steel does not have sufficient cover, the concrete will not fully surround the reinforcing steel with a suitable concrete mix. In this latter case, the concrete section is susceptible to the concrete's breaking loose from the face of the steel; this destroys the bond between the steel and the concrete thereby leading to eventual failure of the concrete structure. Recommended clearances between the reinforcing steel and the nearest face of the concrete cover are as follows: 1 inch for slabs and walls, 1½ inches for beams (stirrups can be 1 inch clearance), and 3 inches for footings. If greater clearances for the reinforcing steel are desired, the designer should be notified so that the engineering can be recalculated. Increasing cover will require a thicker concrete section and more reinforcing. One critical item of reinforcing is the installation of required stirrups. Stirrups are pieces of reinforcing looped around the beam reinforcing and spaced at varying intervals at each end of each beam span between columns or columns and walls. Failure to provide stirrups when specified by the engineer or their improper placement can result in beam failure through shear. This is when the beam slides past itself at its point of support. Shear failure is identifiable by the development of diagonal cracks across the vertical face at each end of the beam span. Underground structures are particularly vulnerable to shear failure because of the heavy sustained loads. Above ground structures may or may not be subjected to their full live load so an error in stirrup placement could possibly never show up. This is not so with underground structures for they are tested to their maximum immediately upon placement of the cover soil. Any vehicular traffic during the cover placement tests the structure even more.

It is said that the underground structure endures its worst load condition at the beginning of its life, but errors in concrete placement can show up months or years later because of material fatigue. Correct amounts and placement of reinforcing steel is critical because a medium sized dwelling with 2 feet of earth cover will be supporting approximately 1 million pounds of soil and concrete. Given this, it is a good investment to have the designer look over the reinforcing prior to placement of concrete. Not to make this small investment in the designer's time is to be penny wise and pound foolish—in the case of an underground dwelling, a million pounds foolish.

Retaining walls are essential with underground dwellings that are open at ground level on one or two sides. It is very tempting to cantilever retaining walls off the concrete shell, but considering the fact that they are often 13 feet high above their footings, after the backfill is placed against them, the stress is too great for the shell and cracks will develop at the point of juncture. It is recommended that retaining walls be self-supporting by utilizing accepted engineering standards. Since the retaining wall is attached to the shell, it is necessary to protect it from movement due to freezing and dry weather. This can be accomplished by extending the footing to a depth of 3 feet below grade. This is a minimum if good bearing is not achieved at that depth, then the footing should be extended to a depth where suitable bearing can be achieved. The importance of minimizing the lateral stress against the retaining wall cannot be overemphasized. This can be achieved by utilizing good drainage through and behind the wall. It is recommended that through drainage be installed every 4 feet with 4 inch tile. Drainage behind the wall can be accomplished by installing tile along its length. In each case, gravel graded from 1½ inches to 2 inches should be placed so that water can filter through to be drained away. Since these retaining walls serve as a transition in grade levels, they generally slope down to the ends away from the face of the shell. A savings in material can be achieved by reducing the size of the footing and the amount of reinforcing as the wall is reduced in height.

Because of the unusually high sustained load that the roof of an underground dwelling is subjected to, it is advisable to allow the concrete to reach its full strength before soil is put in place against it. This strength is not reached until the 28th day. Also prior to backfilling, it is necessary to place crushed rock over the drain tiles and to compact 1 foot or more of fill material over the crushed rock. After this is done the fill can be placed ready for natural settlement. When backfill is not compacted but settles naturally, a certain amount of dressing is required from time to time. It is important that this dressing is not neglected to the extent that ponding of water in settled areas is allowed. Ponding will cause a greater amount of water penetration than normal through to the foundation drain tile. This extra drainage could cause early fouling of this vital part of the building. Only select fill material should be used for backfilling. Fill material that contains roots and sizable rocks tends to prevent achievement of acceptable density from natural settlement. This leaves the roof cover vulnerable to water penetration with no early self-correction.

In conclusion, as with any other condition of life, following directions developed from the mistakes of others will result in a more satisfactory product. All of the items described in this paper are drawn from actual instances observed during nearly a quarter century in the field of engineering. Some will want to invent the wheel again and will not be satisfied that anything is true until they try the system and fail. This paper is not for them, but it is for those who want to build on that which is already learned.

Economic Design Considerations for an Underground House

Lyle D. Feisel
Department Head
Department of Electrical Engineering
South Dakota School of Mines and Technology
Rapid City, South Dakota

OVERVIEW

This paper presents information about an underground house designed and currently occupied by the author. The house is located about five miles west of Rapid City, South Dakota and has been carefully designed to account for the human factor as well as to conserve energy. Advantages and disadvantages of some of the more unique features are considered, along with suggestions which have resulted from the experience gained.

The details of construction of the house are provided. The house has a 1356 square foot interior and is essentially a shell of poured reinforced concrete with fairly typical internal wood frame construction. The house is entirely underground with the exception of two sections of exposed wall. The 15' x 25' living room contains a fireplace and two windows in a 24-foot section of exposed wall facing to the southeast. The master bedroom also has a fireplace and one window in the 16-foot section of exposed wall facing to the south. A detailed breakdown of the construction costs is presented and comparison is made to construction costs encountered in a conventional home.

Temperature measurements are being made using thermistors inside the structure and at various points in the soil outside the concrete walls and in the soil covering the concrete roof. Total energy usage in the house is also being measured. The soil temperatures and energy usage have been monitored since December, 1977. These data are presented and analyzed.

INTRODUCTION

The concept of living underground is as old as Mankind. Nearly every civilization has, at some time and at some stage in its development, employed underground space to provide homes which are secure, easy to maintain, and relatively comfortable. The ready availability of cheap energy, however, directed attention away from underground housing. More recently, with the realization that our energy supplies are dwindling, interest in the underground house is re-awakening.

Many articles, and a few books[1-4] have treated the subject of underground housing. Specific data on such houses have been lacking, however. In this paper, an attempt is made to deal specifically with the performance of an underground house and the economic considerations associated with its construction.

While construction costs vary throughout the country, it will be seen that, at least in this one instance, costs of underground construction are comparable to those associated with conventional methods. Further, a simple life cycle cost analysis will demonstrate that long term economic benefits will likely be quite significant.

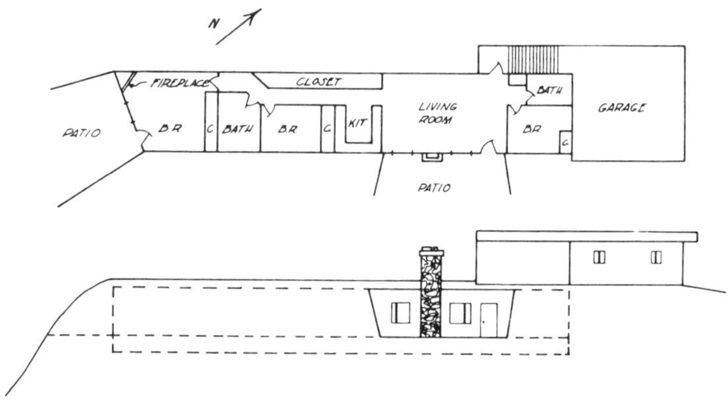

Figure 1. Plan and elevation views of the house.

PHYSICAL CONSIDERATIONS

The house is rectangular in shape with total interior floor space of 1,356 square feet. The floor plan and outside front view are shown in Figure 1. The living space consists of three bedrooms, two baths, kitchen, and living room. There are two fireplaces, one in the living room and the other in the master bedroom. The garage is above ground and is connected to the house by an interior stairway.

The house is completely underground except for a 24-foot wall section at the front of the living room and a 16-foot wall section of the master bedroom at the end of the house. The 24-foot section provides space for two windows each four feet square and a door which opens onto a large patio. The uncovered 16-foot section at the end of the house provides for a third window and a door opening onto the second patio. These windows and patios provide spectacular views across a canyon. The roof of the house is covered with about 18 inches of soil.

Construction details of the house are shown in Figure 2. The structural walls are steel-reinforced poured concrete and are 12 feet high. The walls support 35-lb. steel I-beams on 4-foot centers which in turn support a 6-inch steel-reinforced concrete slab serving as the roof. The walls are set on 5-foot poured concrete footings. Footing drains are provided to eliminate seepage between the floor and walls. Roof and wall joints include a cast-in-place rubber seal to insure water tightness.

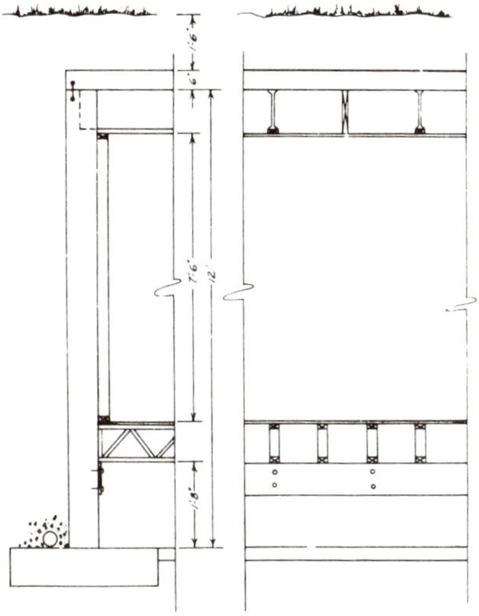

Figure 2. Construction detail.

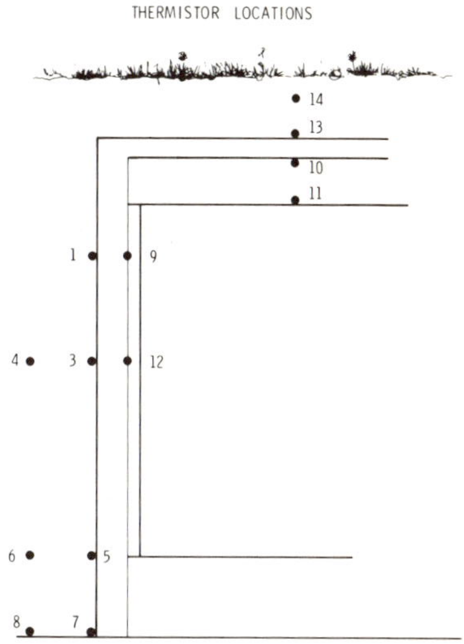

Figure 3. Thermistor locations.

The two-car garage is above ground and is of typical frame construction. The garage is unheated. Laundry facilities are conveniently located in the larger of the two bathrooms.

The internal construction of the house is quite conventional. The floor is supported by 2 x 12 joists set on 16-inch centers. The crawl space under the floor provides space for, and easy access to, all plumbing, electrical wiring, and air ducts. The underside of the floor is insulated with standard batt insulation. The interior walls are 2x4 frame construction with exterior vapor barrier, batt insulation, and sheetrock. Interior partitions are of standard 2x4 construction. The exposed wall sections use 2x6 studs with six inches of insulation. The ceiling is constructed of sheetrock nailed to 2x4's which are bolted to the bottom of the supporting I-beams. The area above the sheetrock is filled with 13 inches of blown cellulose insulation.

To minimize heat losses the windows are triple glazed and the doors are insulated steel with magnetic weather stripping. All floors are carpeted except in the kitchen and bathrooms.

The heating system consists of electric baseboard heaters with individual thermostats in each room. Supplemental heat is furnished by the two fireplaces. Exhaust vents are provided in the kitchen and bathrooms. Hot water is provided by a solar water heater with electric back-up.

INSTRUMENTATION

An instrumentation system has been installed to provide sufficient data to develop and validate a thermal model of the house and surrounding earth. The large thermal time constants of the earth will require collection of data over long periods of time.

Current instrumentation consists of 15 thermistors which were calibrated and put in place during construction of the house. Four of these thermistors are located at depths of 4, 7, 11, and 14 feet below the surface on the back wall of the house. Four additional thermistors are located at these same depths but are buried in the earth 2 feet out from the wall. Two more thermistors are on the inside of the back wall opposite those on the outside at depths of 4 and 7 feet. One thermistor is located on the underside of the concrete roof slab and another just above the sheetrock on the ceiling. Three more thermistors are placed as follows: one at the outer surface of the roof slab, one 6 inches below the surface above the roof slab, and another above the earth surface to measure ambient temperature. Thermistor placement is shown in Figure 3.

An elapsed time meter is installed on the electric water heater. Total electric power input to the house is determined from a watt-hour meter. A modest data acquisition system has been developed using a microcomputer to control the collection of data. The system records data at 15-minute intervals on cassette tape using a modem and tape deck. The data can then be processed on a general purpose digital computer. Funding for the data acquisition system was provided by the South Dakota Housing Development Authority. Technical details of the system are covered elsewhere.[5]

Data collected by the computer system and by manual techniques have been integrated into a data base computer file. The file records experiment day (number of days from first reading), hour of reading, and the temperatures recorded at the fifteen thermistor locations. Due to periodic equipment outages and other factors, various gaps occur in the data. Since data have been taken for two years and since the system time constants are so long, these gaps impose no real limitations on the utility of the data base.

Ultimately, it is anticipated that the accumulated data will serve as a validation basis for thermal modeling studies of the house. As a first step in utilizing the data, however, a computer program has been developed to plot up to six temperatures on any scale between any two times. A continuous plot is provided by interpolation between a periodic data point. This allows a researcher (or the merely curious) to simultaneously investigate temporal and spatial temperature behavior with a minimum of effort. The following section of this paper will be illustrated with plots generated by this program.

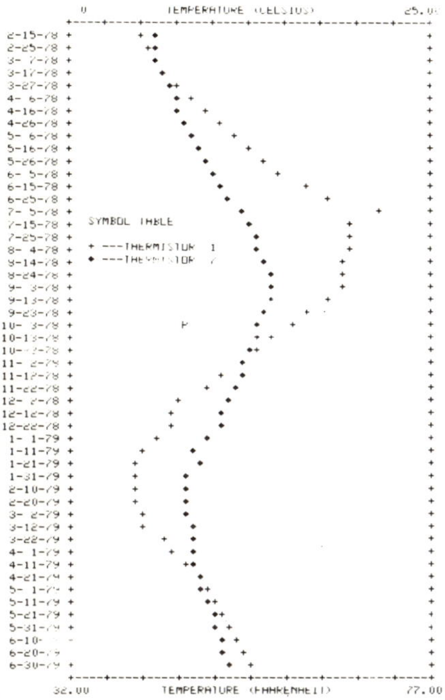

Figure 4. Temperature vs. time. Thermistor 1 (four feet deep) and thermistor 7 (fourteen feet deep). Both thermistors on the wall.

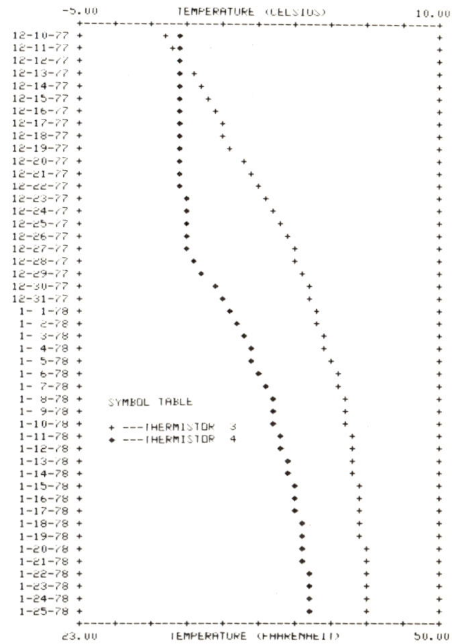

Figure 5. Temperature vs. time for the first fifty days after backfilling. Thermistor 3 (on the wall) and thermistor 4 (two feet from the wall in earth). Both thermistors at a depth of seven feet.

PERFORMANCE

The fundamental principle of earth sheltered construction is illustrated in Figure 4. This is a plot of soil temperatures on the soil side of the concrete wall at depths of 4 and 14 feet below the surface. While thermistor number 1 varies between a winter low of 4 C (39 F) to a summer high of 21 C (70 F), thermistor number 7 only varies between 8 C (46 F) and 13 C (55 F). These compare to air temperature extremes of about -30 C (-22 F) to 40 C (104 F). The moderating influence of the earth, becoming more pronounced at greater depths, is obvious.

Another interesting example of the great heat capacity of soil is illustrated in Figure 5. This shows the temperatures at thermistors 3 and 4 for the first 50 days after initial backfilling. These thermistors are at the same depth, but number 3 is against the concrete wall and number 4 is two feet out into the earth. Backfilling was performed on December 10. By this time the earth in South Dakota, including the exposed walls of the excavation, is well frozen. Figure 5 shows that the temperature of the concrete wall started to increase immediately while the temperature in the frozen earth remained about constant at the freezing point of water for a full 20 days. At that time, it also started to increase with a time constant very similar to that of the wall. Apparently, there was sufficient frozen water around thermistor 4 for the latent heat of fusion to keep the temperature constant for almost three weeks.

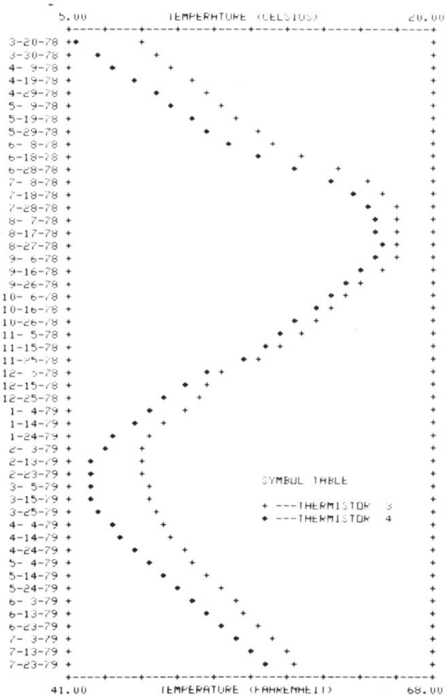

Figure 6. Long-term temperature variations. Thermistor 3 (on the wall) and thermistor 4 (two feet from the wall in earth). Both thermistors at a depth of seven feet.

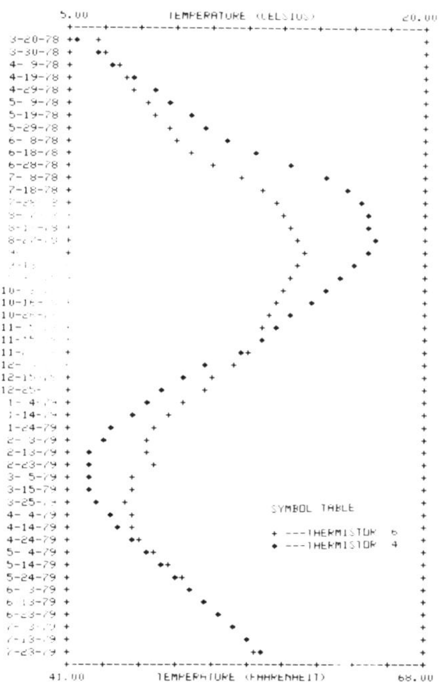

Figure 7. Temperature vs. time. Thermistor 4 (seven feet deep) and thermistor 6 (eleven feet deep). Both thermistors two feet from the wall in earth.

The long-term behavior of these two thermistors is also interesting. Figure 6 shows the two temperatures from March 20, 1978, to June 23, 1979. At no time during the year did the wall temperature fall below the earth temperature two feet away. This indicates that the earth is always a heat sink for the house. The closer spacing of the curves in the period of decreasing temperature indicates that the heat flow to the earth is less during the heating season (descending curve) than it is during the cooling season. This, of course, is the optimum condition for both heating and cooling.

Vertical heat flow is illustrated in Figure 7 which shows the temperatures of thermistors 4 and 6 located two feet from the wall at depths of 7 and 11 feet. The crossover points (indicating zero vertical heat flux) occur about April 15 and November 15. Thermal lag is illustrated by occurrence of the minima and maxima at the deep thermistor (6) about one month later than at the other thermistor which is 4 feet higher.

While the deep temperatures exhibit excellent stability, temperatures at points on and above the roof slab are less stable. Figure 8 shows a plot of the temperatures of termistors 10, 13 and 14. These are located on the inside of the roof slab, on the outside of the roof slab, and one foot above the roof slab (i.e. six inches below the surface). It is apparent that these temperatures are far more sensitive to ambient temperature. Even here, however, the variations of ambient temperature are smoothed considerably.

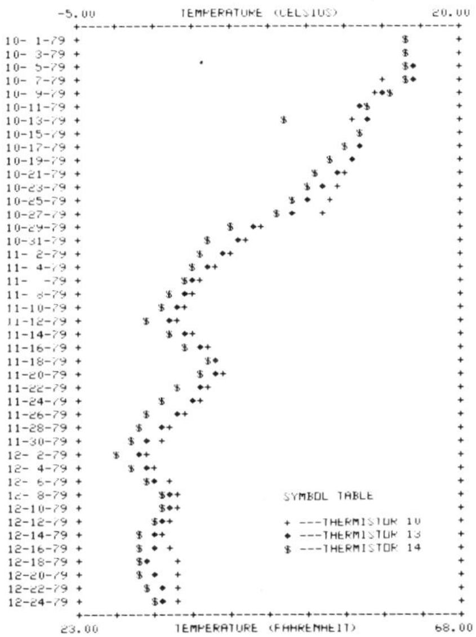

Figure 8. Temperature vs. time. Thermistor 10 (bottom surface of roof slab). Thermistor 13 (top surface of roof slab), and thermistor 14 (one foot above roof slab or six inches below the earth surface.

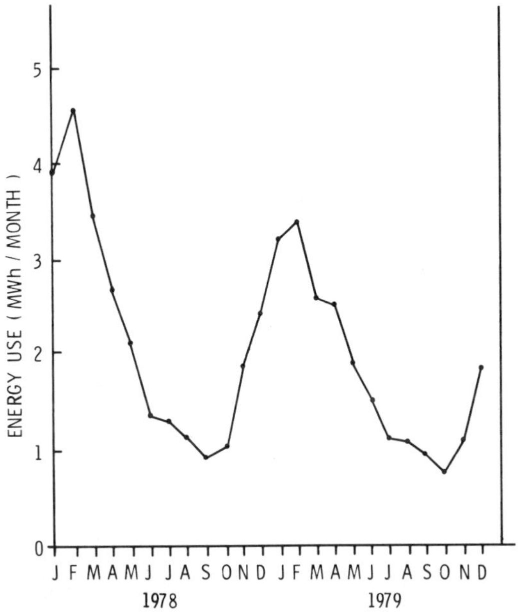

Figure 9. Energy use vs. time.

The temperatures recorded in the data base contribute to a qualitative understanding of heat flows in an underground house. As a mathematical thermal model of the house is developed, the data will allow a quantitative calibration and validation of the model which will then provide design data for further earth sheltered development.

The equipment will also be used in the future to obtain system step-response data. The instrumentation System will be expanded to record individual room temperatures. Then, with the house unoccupied, all energy inputs will be removed and temperatures will be recorded for a period of three days. In conjunction with the model this will provide system time constants and, possibly, thermal properties of the surrounding earth system.

ENERGY USE

The electrical energy used in the house is plotted in Figure 9. This plot shows energy used during each one month period centered about the first of the month in question (i.e., from the 16th of one month to the 15th of the next). The plot has several interesting features.

The initial heat-up of the house is apparent. The house was first occupied on December 17, 1977, and it took a few days to get all heaters and thermostats working properly. Hence the January usage was lower than that of February. February usage (January 16 to February 15) was highest since this was the period of coldest outside temperatures and the surrounding earth was still being heated. Also, during this first winter the backfill had been accomplished to the level of the roof but the roof itself had not been covered.

The second winter of occupancy is very similar to the first but with lower use in a month-by-month comparison. Usage still peaks in February but it peaks at about 3.4 MWhrs as opposed to 4.5. Data from the winter of 1979-1980 will provide further information on the long-term stabilization of the house's thermal environment.

Because of different family characteristics, different lifestyles, etc., it is difficult to make meaningful comparisons between energy uses in different houses. As a best effort, energy use is analyzed on the monthly basis in terms of Btu/ft^2/degree- day. Since English units are most prevalent in this particular measurement, they are used here.

The temporal aspects of the analysis are illustrated in Figure 10. This figure shows adjusted power usage (energy used during each one-month period minus the base summer load of 921 kWhr) and heating degree-days per month plotted on the same time axis. Again the large thermal lag is apparent from the delayed occurrence of the peak of the power curve relative to the peak of the degree-days curve. After the weather has started to moderate, the heating system is still replacing heat taken from the earth during the colder weather.

While these curves are interesting, summary results are probably more useful. During the 1978-79 heating season, the house required total heating energy of about 40 million Btu's. This converts to about 28,000 Btu/ft^2 or less than 3.5 Btu/ft^2/degree-day (8144 degree-days[6]). Again, one must exercise caution in making comparisons, but typical frame construction houses generally require 10 to 12 Btu/ft^2/degree-day. That is, 40,000 to 45,000 Btu/hr for a 1200 ft^2 house on a 75 degree-day day.

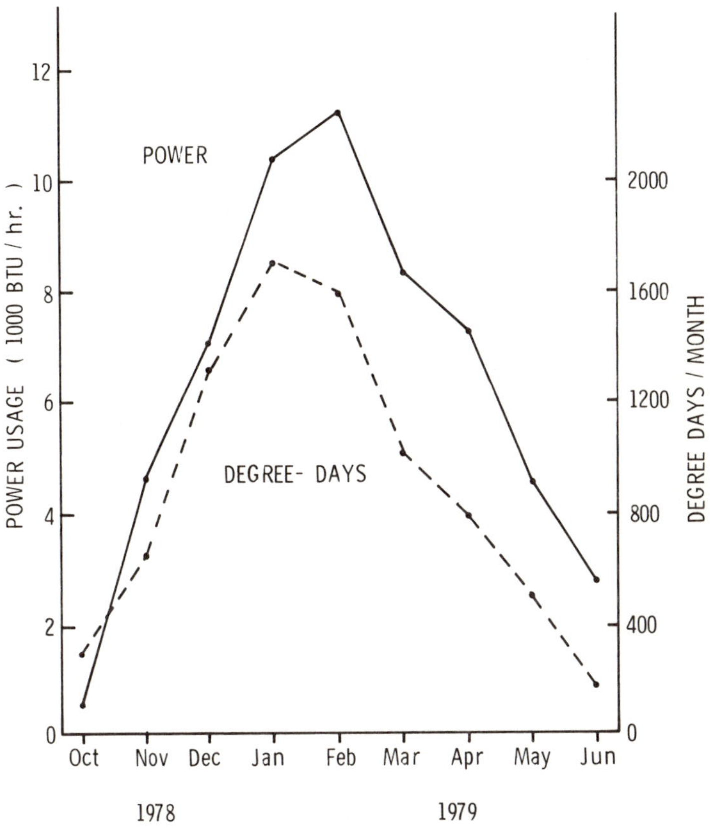

Figure 10. Power input and heating requirements vs. time.

In summary, the energy use of the house is on the order of one-third the use of a comparable frame dwelling. This can result in substantial savings to the homeowner. An additional, less obvious, advantage exists in the thermal lag that is a natural consequence of earth-sheltering. Energy consumption is very insensitive to temperature fluctuations extending over periods of several days. This property is of great advantage in solar heating (both passive and active) and is of considerable significance to an energy utility in load levelling.

ECONOMIC CONSIDERATIONS

While there have been some interesting discussions about the economics of underground construction, and some rather detailed predictions have been made, the fact remains that we have no real experience in this area. There are still questions about energy savings, maintenance costs, and, especially, the resale value of underground houses. An attempt will be made to investigate some of the parameters associated with the economic analysis, but no firm predictions will be made here.

The analysis will be broken into two parts: the cost of construction and life cycle analysis.

CONSTRUCTION COST

An interesting feature of underground homes is that the house area is generally computed on the basis of the inside dimensions. This is to be expected since the outside of the walls is buried and, therefore, inaccessible for measurements. In conventional construction, however, floor area is computed on the basis of outside foundation measurements. In order to achieve an effective comparison, the area of the underground house was adjusted to compare to a similar inside area in a conventional house. The inside area of the underground house is 1356 square feet, which is the area of a square room 36.8 feet on a side. If this square room were built by conventional techniques, the outside dimension would be 1 foot larger in each direction, or 37.8 feet square. In conventional terms, then, this room would be said to have an area of 1431 square feet. This is the number used in the calculation of costs.

Construction costs vary throughout the country. For the purpose of comparison, builders' estimates in the Rapid City, South Dakota area range from $35-45 per square foot. This is considered quality construction, but does not result in a house with many luxuries or with finer points, such as hardwood woodwork, etc.

The costs associated with the construction of the underground house are summarized in Table 1. They are broken into two areas' the cost of the concrete and steel shell and the cost of the interior.

Turning first to the shell cost, it is apparent that the excavation costs were relatively low. This figure includes excavation of the original site, installation of a septic tank and drain field, burial of electrical and water service, and two different stages of back-filling and covering. No blasting was required in

COST OF SHELL:	
Excavation	$2,830
Concrete	12,880
Steel	7,950
Labor	1,000
Builder's Fee (pro-rated)	1,000
	$25,660
COST OF INTERIOR	
Building Materials	$10,310
Insulation	1,180
Plumbing	2,420
Electrical	2,120
Fireplaces	1,270
Cabinets	2,190
Carpet	2,530
Labor	7,490
Builder's Fee	3,000
	$32,510
TOTAL COST	$58,170

Table 1. Construction Costs.

the excavation process, but much of the digging had to be done through weathered limestone. The amount listed for concrete includes not only the concrete mix itself, but the labor associated with the setting and stripping of forms and the reinforcement bar installed in the concrete.

The builder set the steel I-beams and poured the roof deck. The labor costs and builder's fee for this function are estimated. The total cost of the shell, then, is in the vicinity of $25,640.

The items included in the interior cost are quite conventional. The materials package is rather small compared to a conventional home because of no need for shingles, siding, a large number of windows, etc. Finishing of the interior—including painting and carpeting—is about $32,510, for a total cost of $58,150 for the structure.

Using the effective area of 1431 square feet, this yields a building cost of $40.64 per square foot. Clearly, this is competitive with the conventional construction method in use in this area.

There are, of course, additional costs associated with the total construction project. They include such things as septic tank, driveways, patios, a surface garage, and sidewalks. Since these costs are site specific and associated with site development, they have not been included in the costs of the structure itself.

LIFE CYCLE ANALYSIS

The previous section has shown that the construction costs of an underground house are comparable to those of a conventional structure. This section investigates the thesis that a builder can actually afford to pay more for an underground structure if the total life cycle costs are considered.

Life cycle cost analysis is based on the concept that an investor can invest money at the front end of a project if there are resultant savings in operating costs.

It is obviously a predictive process and—at least with regard to underground residences—we do lack experience. This dictates the use of a conservative approach to the analysis. While rather elaborate processes of life cycle cost analysis have been developed, a very simple analysis will be performed here. The purpose is more to illustrate the rationale of life cycle cost analysis than to make specific predictions as to savings of this type of construction.

The analysis is performed under several quite conservative assumptions. One: the house has a usable lifetime in its present form of 30 years. Two: there will be no inflation for the next 30 years. Three: there will be no escalation of energy costs during that period. And four: an interest rate of 10% will be maintained over the next 30 years.

It is difficult at this point to predict the exact savings that might be expected from the underground house as compared to the conventional structure. It is estimated, however, that annual maintenance costs will be cut from \$500 to \$250 and that the annual energy use will be cut from \$1500 to \$750. This results in a total savings of \$1000 per year.

The question, then, is what is the present value of an investment which will save the investor \$1000 per year; or, stated in a somewhat different fashion, how much money can be borrowed now and paid off in 30 years at \$1000 per year and 10% interest. This is a straightforward direct reduction loan situation, and the principal amount is computed to be \$9,427. Thus an underground structure may be considered to be worth about \$10,000 more to its owner on the day that it is occupied than a conventional structure of similar size.

This simple analysis has been performed subject to very conservative assumptions. The effect of inflation, of course, will tend to further increase the present value, since it will be easier to pay off the loan with inflated dollars. Also, the effect of fuel cost escalation—which is occurring at an astounding rate—will cause further life cycle cost advantages. If fuel costs continue to escalate at their current rate, the \$750 per year savings will be conservative indeed.

Another aspect of life cycle cost analysis, which is even more difficult to assess due to our complete lack of experience, is the phenomenon of salvage value. Today, a house built using conventional techniques lasts 50 or more years with negligible salvage value. As the house reaches the end of its life, it is generally just destroyed and a new house built on the site. An underground home, on the contrary, may have a similar lifetime for the conventional part of its construction, but lifetime of the steel and concrete shell is predicted to be 300-500 years. Thus, when an underground house reaches the end of its first lifetime, its salvage value is not zero, but is rather equal to the cost of the shell itself. Referring back to Table 1, we see that is $25,640 — or approaching one-half of the initial outlay for the house. It will, of course, be several decades before this hypothesis is tested and before one will be able to observe the effect of this increased salvage value on house resale costs. It is anticipated, however, that it will be significant.

The economic considerations of underground construction may be summarized by saying that underground construction costs are comparable to those associated with conventional methods and that life cycle cost analysis would justify spending considerably more.

CONCLUSION

It has long been recognized that underground housing will result in energy savings. There has been, however, some concern over the psychological aspects of life underground and over the costs associated with the construction of the house.

To date, the residents of the home discussed here have found the underground to be a very hospitable environment both physically and psychologically. It is comfortable, quiet, and secure—three attirbutes which are difficult to achieve in modern American life.

Construction costs were found to be approximately the same as those associated with conventional methods. Furthermore, life cycle cost analysis assigns a significantly higher present value to an underground house due to the reduced maintenance and energy costs throughout the lifetime of the house. Consideration of ultimate lifetime or salvage value further enhances this effect.

Finally, energy use in the house has proven to be significantly less than in a conventional structure of comparable size. An additional benefit is the large thermal lag which significantly increases the effectiveness of both passive and active solar heating systems.

As the psychological and economic benefits of underground housing become better known, and as energy costs increase, this type of residential construction is certain to become more significant in the United States and abroad.

REFERENCES

1. *Earth Sheltered Housing Design*, Underground Space Center, University of Minnesota, Minneapolis, MN

2. *The Use of Earth Covered Buildings*, Frank Moreland, Editor, Superintendent of Documents, U. S. Government Printing Office, Washington, DC 20402

3. *Underground Designs*, Malcolm Wells, Box 183, Cherry Hill, NJ 08002

4. *Earth Integrated Architecture*, James W. Scalise, Editor, The Architecture Foundation, College of Architecture, Arizona State University, Tempe, AZ

5. Sletten, Per S., *A Microprocessor Data Acquisition System*, unpublished M.S. Thesis, Devereaux Library, South Dakota School of Mines and Technology, Rapid City, South Dakota.

6. *Local Climatological Data*, National Oceanic and Atmospheric Administration, Rapid City, South Dakota.

Earth Contact Building: Accent on Post-Tensioning of Concrete*

Lon B. Simmons

Simmons & Sun, Inc.
P.O. Box 1497
High Ridge Shopping Center
High Ridge, Missouri 63049

Deservedly, there is today a great hue and cry being made about improving the efficiency of our housing and commercial buildings. It can be safely stated—to most people's amazement and disbelief—that the quintessential energy efficient and environmentally compatible building can be built only by going underground. The natural isolation provided by being underground allows the discerning designer to epitomize efficiency.

Despite the fact that going under showed great promise, I rejected the idea of building the houses out of conventional reinforced concrete. The inherent tendency of concrete is to crack. Everybody knows that. The molecular structure of concrete is such that it is most conducive to cracking.

This is not to say that all undergrounds will crack and leak, but the possibility is great. I personally feel that concrete block and dirt over wood is an open invitation to disaster. As a builder seeking to establish some credibility in the field, I could forsee the possibility that, if the concrete failed somewhere down the line, my name would be mud, both literally and figuratively.

So I set out to find a better way to build. I believe that I have found it. It is a technique called post-tensioned concrete, which has had thirty years of proven commercial application. It has been used very successfully in large concrete projects, both above ground and below ground, where strength and water-tightness is essential.

The St. Louis Gateway Arch, airplane hangers in Kansas City with a 330 foot clear span, countless parking garages, concrete water towers, and skyscrapers in areas prone to earthquakes are all supported by post-tensioned concrete.

Basically stated, post-tensioning cables replace the reinforcing rods in concrete. In conventional construction, reinforcing rods work only when the concrete fails. For example, witness the cracks that most people have seen in basement walls. The concrete cracks, but the reinforcing rods hold. However, there is still a crack and, often, a leak.

In the post-tensioning method I use, the concrete is poured over a grid of ½ inch seven-strand steel cables. The cables are encased in a thin nylon covering with a lubricant inside which allows the cables to slide once the

**This paper has been previously published in, and appears here with the permission of,*

concrete is poured around them. One end is inserted into an anchor which is placed just inside the form on the floor slab. The cables run the full length of the slab and extend through another anchor in the opposite form.

After the concrete is poured and reaches a certain strength, a hydraulic jack is attached to the exposed cable and pulled to 80% of what it takes to break the cable. When the jack is released, the nosepiece drives two wedges around the cable which extends through a tapered hole in the anchor. Once the jack is released and the wedges lock the cable, 60% tension on the cable remains. The exposed end is cut off and the anchor is patched with cement so that no part of the cable is exposed. The pulling of these cables compresses the concrete, making the concrete stronger and watertight. Imagine interlocking your fingers, representing the tendons, and pulling your fingers very tight. This is exactly the process one uses when hoisting a friend up over a wall. Superior strength is created.

Also vertical tendons are placed every 4 feet before the concrete is poured in the floor slab. After the slab is poured, standard tendons run horizontally through the walls. The walls are then poured around the tendons and the horizontals are pulled. In the roof deck the tendons are placed much closer, averaging 12 to 14 inches apart, depending upon the span desired.

After the roof tendons are pulled, then the vertical tendons, extending from the footing up through the walls and roof, are pulled. Simply stated, a large box is built and pulled together. The cables extend from the side walls and through the wing walls and are pulled together.

Since the south side of the house is exposed for solar access, a 4 foot retaining wall is poured on the roof deck to hold the 3½ feet of dirt on the roof. Vertical tendons on the south side begin in the footings, extend up through the wall, roof, and retaining wall, and then are vertically post-tensioned.

Besides creating a superior structure, the advantages of post-tensioning are numerous.

First, there is a considerable savings on steel. In a 3,200 sq. ft. structure, less than 6 tons of steel are used. This includes the post-tensioning cables and a series of ½ inch reinforcing rods that are used in the roof. No steel beams are used, and the largest piece of steel is ½ inch in diameter. A contractor pouring for me in Illinois used 35 tons of steel to build a comparably sized underground house last year. Any one knowledgeable of the price of a ton of steel added to the installation cost is aware of the savings experienced here.

Too, less concrete is used. In spans up to 22 feet, we pour a 6 inch roof deck. I have seen contractors in conventional construction use 10 to 18 inch roof decks. Each additional inch of concrete across some of my houses requires 15 cubic yards of concrete.

Also, a considerable savings is realized in waterproofing. The post-tensioning method has been used in million gallon water tanks with no waterproofing on the inside. In one such water tower, the walls, in conventional concrete construction, would have to have been 2 feet thick with tons of 2 inch reinforcing on the inside to hold the tremendous pressure generated by a million gallons of water.

In this particular water tower, using ½ inch steel cables, the walls are 9 inches thick. I am proceeding from the premise that, if indeed this structure can hold a million gallons of water inside post-tensioned concrete, they can keep a few drops of rain on the outside from getting inside my houses.

We are able to waterproof the structure at an approximate cost of 5 cents per square foot. Many people maintain no waterproofing at all is needed. However, just to be sure and to create a vapor barrier, we apply a waterproofing membrane.

Drain tile is installed around the footings to drain water. The month of April 1979, saw the St. Louis area receive rain sixteen days out of thirty, including one of 6 inches. We had one exposed structure ready for backfill that took all this rain and showed no signs of leaking.

Another advantage of post-tensioning is its superior rigidity. In backfilling the houses, the most deflection I have experienced is less than ½ inch. In most cases no measurable deflection is experienced. By going to a 10 inch deck and the addition of more tendons, we are able to clear span up to 30 feet with no interior bearing walls, even with 3½ feet of dirt on the roof. In these large spans, the engineers calculate the amount of deflection and have the tendons draped in such a way as to raise (camber) the roof when the tendons are pulled. Then when the load is applied, the roof becomes flat.

A special word must be said about the insulation, which is critical on an underground house. Most people are acquainted with R-factors, the resistance of a substance to conduct heat. The desired R-factor in walls and roofs will vary in different parts of the country.

In Missouri, an R-30 roof and R-20 walls are desirable. I see many people make the critical mistake of building a concrete structure and backfilling it with dirt. Contrary to what most people think, dirt is not a good insulator. It is, however, a very good isolater.

Concrete is also a very poor insulator; it has an R-factor of .08 per inch. Therefore, a 8 inch standard wall gives one an R-factor of .64. Thus, when people build concrete structures and backfill them with dirt, and attempt to heat it, they are attempting to heat the surrounding mass of concrete and earth. This rapidly becomes a losing battle. However, concrete has a fairly good storage capacity. Insulating on the interior of the concrete ignores a basic fact of physics in that concrete has a specific heat of .22. Water, the most perfect natural storage medium, has a specific heat of one.

This means that it takes roughly five times as much concrete as it does water for storage. A passive system depends on mass for storage. Therefore, if one moves the insulation to the outside and insulates the exterior, one is able to use the areas of concrete which are insulated for storage.

In some houses, depending on the amount of concrete used, we have a storage capacity equivalent to 20,000 gallons of water. The thicker insulation should be on the roof as the heat loss is greatest through the roof and the top parts of the walls. I use 6 inches of rigid insulation on the top with decreasing amounts going down the side wall.

The floor slab is left uninsulated to favor natural cooling. On the exposed south side, a frost wall is dug three feet down, lined with two inches of insulation and 10 inches of concrete is then poured back behind it. This prevents extreme temperatures from coming up underneath the floor.

Another tremendous advantage of underground houses is that they lend themselves perfectly to solar utilization. While active systems may be needed for the massive job of converting existing homes into solar, in addition to many other uses, in new home construction, if proper siting can be achieved, the passive system is the more viable alternative.

There are several ways to achieve passive solar heating in an underground house, whether by use of Trombe walls, direct gain or central atriums. The houses have no air conditioning system; they are 80% solar heated. The only backup system in most cases is a wood burning stove. However, we have some east-west slopes; we have houses facing north where the land dictates that the houses cannot be faced south.

These systems require conventional heating and air conditioning systems, although they are very small in Btu output and will rarely have to be used. They are not advertised to be passive solar, but rather energy efficient buildings.

In some cases, a Trombe wall is used on the south side. It is a concrete wall 15 inches thick, painted black on the outside covered with double pane tempered glass 6 inches away. As the sun shines in the winter on a low angle, heat is generated in the 6 inch air space. On a cold day in Missouri with the sun shining and a temperature of 0 F outside, the outside temperature of this wall can hit 130 to 140 F. Vents are placed at the top and bottom of the wall. Hot air naturally rises through the top vents. As this heat source is created, cool air is drawn through the building in through the bottom vents. A natural thermosiphoning action is created without benefits of pumps, fans, motors or thermostats. Heat radiates through the wall about two o'clock in the morning at about 90 F. The building uses the three principal methods of heating, conduction, convection and radiation, and the building is primarily heated by radiation.

The amount of heat stored in the Trombe wall and the enclosed concrete in the building is sufficient to hold a house for a week of sunless weather with only a 4 to 6 degree drop in temperature if no other supplementary heating is used. During periods of sunless days, a wood burning stove is used.

Two mechanical systems are used inside the house: both are very efficient fans. One is for use in the summer, one for the winter. One fan is placed on the wall between the house and the garage. It exhausts into the garage and when the windows are open on the south side, it is possible to move 4500 to 7000 cu ft of air per minute.

The other fan is used in the winter and is located in the living room near the wood burning stove. During periods of sunless days, warm air generated by the wood burning stove is pulled from the living room and ducted to each room in the house. In most cases, only 20,000 Btu are required to heat the house. This is easily generated by a quality wood burning stove.

It should be noted that one of the advantages of the passive approach is its simplicity. I have been very careful to design houses that can be built anywhere in the United States by any competent contractor. The passive approach requires no breakthrough in technology.

Simply stated, it is taking standard building components, adjusting them to climatic and structural conditions, incorporating and maximizing the physics of the structure and mixing in a generous portion of common sense, an ingredient which, I humbly submit, seems to have escaped the building industry thus far.

Designing, Building and Living in Underground Homes

Buck and Janitye Vaile
620 Vancouver Drive
Westerville, Ohio 43081

The company we run is called Solar Earth Builders. We call the houses we build "underground", rather than mince words with all the names for underground buildings. These are simply underground houses, and we are proud of that fact. We call them "Solar Earth" because we combine the assets of the earth with those of the sun. The houses do have passive systems. In fact, we designed our house with two banks of passive inputs. One is the southern face at the lower level, and we have a solar atrium at the upper level which picks up the sun.

I have one partner and one associate. Our background is not what one would expect as normal for the underground home builder. My parents were both professionals who were involved in construction, although it would not appear so. The memories I have or my early childhood and family was that we were always building something. My mother had a fancy for having rooms torn out, redoing and redesigning, etc. As early as I can remember, I was always involved with construction. When I completed professional training, instead of following the field that I trained for, I just couldn't seem to get out of construction. We initially started a company to build swimming pools. We graduated into home remodeling. We built and designed pool enclosures because we couldn't find any enclosures that were doing what we thought they should do. We first built above ground pool enclosures. The enclosures were aimed to pick up and take advantage of the sun. As it turned out, these enclosures were more energy efficient than the houses we attached them to. However, the people who were building underground homes three or four years ago were mostly do-it-yourselfers, thank God. These people had enough guts to go out and try something while everyone else just sat and talked about it.

So, a little over three years ago, we started in Columbus, Ohio, with a design we had worked on for some period of time. It was an underground design, and we felt it was time to stop talking and start doing. We had not previously built an entire house, only some remodel work and pool enclosures (that were almost as big as homes). So really, building underground was our first experience in building an entire home.

We have consulted on numerous other homes. Some people have tried to refer to us as "experts", since we have had active experience in building underground homes. I don't like that term very much. I recall that when I took algebra in the eighth grade, "x" was a symbol for an unknown, and "spurt" according to the dictionary, was a "drip under pressure". I prefer to refer to myself as a learner. Technically, I might be referred to as a thief. Maybe a nicer word would be as a borrower, because when we started in this business we stole or borrowed all the knowledge we could find on underground homes. One of the underground designers whom I highly respect is Malcom Wells.

One of the things that Malcolm Wells has contributed to the underground movement was not just his expertise in showing with his designs how wonderful underground was, but Malcolm Wells did something which I think is very important, which we try to emulate in our business: he told everybody what he did *wrong* not just what he did right. Once you have done something and done it right, that's fine. But the secret in building is to find out what has been done wrong. In his articles, he always stresses the mistakes he has made. Thus, we have been able to avoid making these same mistakes in our homes. Perhaps this is one of the reasons that our house has functioned so well. The first winter we were in the house, we burned approximately one cord of wood. The use of the term "cord" is not a real good measure because a cord of wood depends on someone's idea of the size of a cord and also on what type of wood you have. Thus, when we calculated how much wood we were burning in terms of cords, we tried to weigh the amount of wood. There is a relationship between the weight of the wood and the amount of Btu's produced from that wood. There is not much relationship between the volume of that wood and the number of Btu's, however. We used about 3500 pounds, which would be about a cord of good hard wood, to heat our house that winter. We were quite pleased with that. It is interesting to note that last winter we used a little more wood. We couldn't figure that out because it didn't seem to be as devastatingly cold as the previous winter. We didn't have the blizzard we had the previous winter and we weren't confined inside for 4 or 5 days in a row as we had been the previous year. However, when the reports came back from the national weather service, we found that the mean temperature actually had been colder than that for the winter before. On the basis of degree-days, we used the same amount of wood both years, so the house functioned very nicely.

When we built our house, we did make some mistakes. We have what is called a solar atrium, and we are pleased to see that a number of underground builders have picked up that term. This is really the back of the house where we have solar panels on the first house; these were on a slope of about 30°. On subsequent houses, we have put them on vertical surfaces. One of the problems we've had with that first house, was that there was far too much heat in the summertime, and frankly, not enough in the winter. In the fall and spring it was very nice, because it supplemented the amount of heat we needed for the house quite comfortably. The windows in the southern face, in the winter when the sun was out and when the temperature reached anywhere above 35°, made it basically unnecessary to even use the stove inside the house.

We had shades inside so we can help bring in the sun when it was out and then try in some manner to insulate the glass a little bit better. Obviously, the glass is the area where you lose the most heat, and if you have much more glass in any except southern exposure, you're going to lose much more than you gain. Certainly, the passive system we had via the solar atrium was not sufficient because of the angle. We changed the solar atrium and now it's working a little better. Actually it now doesn't give as much heat in the spring and the fall, but it does give a lot more in the winter; it does not give us an extreme amount of heat in the summer. Previously, about an hour after the sun came up, the solar atrium would get up to about 120° and above. We then had to use a lot of electricity to run our exhaust fan to pull that excess heat out. Then, we changed our house around to make it a little more passive.

Another mistake we made was to install a hot water heating system, an active system as opposed to a passive system. The *active* nature of the system was not the mistake, the placement *was* a mistake. We thought, and it seemed logical at the time, that if you placed the hot water tank close to the panels, the fluid flow would be a very short run and therefore you would get a lot of heat. Unfortunately, the tank wound up in the garage some 60 or 70 feet away from where we are using the water. That doesn't workout well at all. In all subsequent designs, the hot water tank is within about 10 feet of were the hot water is going to be used. We allow the runs to the panels to be a little longer. After all, that's free heat and you don't have to pay for that run, but you do have to pay for the run from tank to fixtures.

We certainly made mistakes in the waterproofing. It is interesting in this business that people are very concerned about waterproofing. Obviously, when you have an underground house and put a couple of feet of dirt over the roof you think, "if I ever have a leak, I've got to go through that dirt and how am I going to find it?" After once sealing them up, we haven't had leaks in the roofs. We have had leaks in other places where we should not have had them. We had a leak at a vent we put up for our stove. We had a leak at a cold joint in an above ground part of the house. This happened in a 4 inch rain where water filled up enough to lie on the roof and come into the aboveground area through one of the seams.

We had some problems with thermal breaks. When you run concrete from underground to the outside, and if you don't "break" that concrete path, you have a bleeding effect. The heat will move into the wall in the underground portion and bleed right through the wall to the outside unless you use a thermal break. That was one of the little tricks or points Malcolm Wells emphasized in his writing. What he didn't mention is that you have to waterproof these thermal breaks or they may act like there is a broken water pipe during a rainstorm and flood the whole house. We had to dig that one out and fix it.

We did not expect to build that house and not make any mistakes. Our goal in building the house was primarily to show that it could be done—we showed that. One of the subsidiary goals was that we wanted to build a house that was contemporary in design, had a lot of light in it, and didn't give people the impression that this was going to feel like a cave. I just say, frankly, if we took somebody into that house or any of our underground homes while blindfolded then removed the blindfold they wouldn't know they were underground. Our homes are well lit. There are many houses not built like ours and I have no intention of saying that those homes are not constructed properly. In fact, in the near future there will be much, much more information coming forth on underground and underground with passive solar than we have begun to touch on. I hope that we can share and help supply some of that information. We built ours for specific purposes with certain goals which we met. Those goals need not be met by everyone. I think it *is* important that people should simply get out and *do* it! Until you do it, you don't know whether it is going to function or not. You can have all the studies and all the bureaucrats sitting around the table, writing reports, deciding how something should be done, however, unless there is someone actually out there to take the bull by the horns and do the actual construction, you'll never know how it functions. We have gone from a situation where we initially used a pre-cast slab or pre-cast planks to where we are using a poured-in-place roof for our homes. One of the reasons is that we like the greater mass. Another is that we like the flexibility of the poured-in-place roof. We think it is stronger. Yet, I don't think that poured-in-place is the end all, at least not the way we are doing it with standard steel reinforcement. I think the best way presently available is through post-tensioning of the house slabs.

We try to make improvements. We have built a home underground that is comfortable. But, we have some other areas we wish to go into. Another concern of ours is water. We are trying to integrate into our homes composting systems and recycling systems. The two homes we are working on now both have composting systems. We're selling surface discharge systems in these homes. We have not gotten to recycling systems for grey water. But, we are saving somewhere between 30 and 50 percent of the normal water usage just with the composting for black water. Hopefully, in the future we can save as much as 80% on the water usage. On the basis of only one or two homes, we are not making much of a dent in the nation's problems, but, if we can learn and others can learn from us—then take our experience a little further—we can help. People can learn from the mistakes we have made and will make in the future. Hopefully, that will make a dent on the problems of this country and those of the world—just through pure water.

We built to set an example. We are continuing to build in order to learn. There are many advantages and there are some disadvantages in regard to underground homes. I suggest that if you are thinking of building underground, you need to consider just *why* you want to build underground. You should write these reasons down and look at them. You should consider whether these reasons are going to be valid a year, five years, 10 years, even 50 years from now. If after careful consideration, the underground home still looks good to *you*, by all means *do it*!

We have found that some of our reasons for building underground have changed. At this point I would never consider building a permanent structure for myself and family that was not underground. I am that convinced. However, my reasons are not the same as they were when I began thinking about building underground years ago. As a builder there is a feeling about the structure I have when I look at the underground home, at least the way we are building them, that makes me feel like it is going to be there a long time and not to worry about the wind blowing it down. Although I didn't build that house because it was tornado-proof, as weather patterns become more severe it does become a consideration. For example, hurricane Frederick dumped about 5 inches of water in an area that included an underground home. Prior to that, the underground home owner had been receiving a lot of razzing for "living in a hole in the ground". After this rain had caused schools to close and all kinds of flooding, this underground homeowner went to work as usual. He noticed that several people who had been razzing him were not at work that day. It turned out that they were at home pumping out their basements. Moral: underground houses built properly are very watertight.

There are two things I would consider if people ask "I want to build underground and what should I look for?" It is essential to either face south or plan very well for what happens when you don't face south. Underground homes are capable of storing a great deal of heat. We built a house facing east because there was a beautiful view. The owners wanted to face the east. That was a mistake, based on the information we have. We have since gone in and corrected some of the things affecting comfort in that house. An east face picks up a fantastic amount of heat in the morning in the summer — and of course a west face in the afternoon. So in the east facing underground house the owners have to be careful in the mornings in the spring, summer, and fall because of the extra heat gain there. The house should face south.

The advantage of building underground, especially in conjunction with passive solar, is energy efficiency. We try to stay away from active solar because, frankly, it is complicated. We like a system that is simple, that is very much in keeping with the environment, that doesn't destroy anything around it and that doesn't require high technology. It may not be the right way, but that is how we are geared and that's where it's at with us. Yes, underground homes are very energy efficient, but I should point out that some above ground—and certainly some passive solar homes—have the potential to be at least as energy efficient. That should not be your only reason for going underground. Underground homes are quiet. Surface homes could be made quiet but those underground are naturally quiet. Again, that shouldn't be the only reason. Underground homes are a little cleaner. You don't feel the temperature changes in the underground as rapidly as you do in the above ground homes.

The following has been published in Earth Shelter Digest and Energy Report, *and appears here with their permission.*

Since 1977 we have had 50 to 70,000 people through our earth sheltered home. I have shown thousands of visitors what I think is, without exception, the only way to live.

The southern exposure all across the front of the home gives me light, naturally, every light hour of the day, all year long.

The way the solar atrium is designed, I have that same southern exposure coming in the back of our home. So, even though I'm 12 feet under, those homes are probably twice as light as most aboveground structures. We have a Jotul wood burning stove. With this, our primary heat backup we have lived through two of the coziest winters I have ever spent in good old Ohio. We burn about 45 pounds of wood a day. Now that would burn up in about 45 minutes in an open fireplace.

On a sun-assist day we use anywhere from 5 to 35 pounds per day. We set the fire with three 12-inch logs that average about 5 pounds each, then close the stove down after a nice burn is started. The reason for this is that this stove was designed for 2000 square feet above ground. We found that in our 2040 square feet underground it simply produced more heat than we could possibly use.

The radiant heat is pulled into a vent at the ceiling and dispersed throughout the home in the duct system. I must say that I have really enjoyed not having any drafts to contend with down here, as I had to above ground.

The bedrooms are both light and airy. As far as your design is concerned, you can have any decor you've always wanted. Our windows open out. And every major room in the house has two fire exits.

The master bedroom may be of special interest to you. It is to us. It was here on February 16, 1979, that we gave birth to the first underground baby. Delivered at 4:08 a.m. by Daddy Buck.

The solar atrium is probably the most fun room in our house. It is an absolute haven for the children. The little ones thought it was "outside" for a long time. (I was not about to tell them differently.)

For a while we had a garden in here. I planted beans, cucumbers, lettuce and tomatoes. Every package I planted said 7-14 days. In here they were all up in 4 days. (However, I learned one very important thing. If it needs to be cross pollinated, you should start it in here then take it outside. I found that I am a terrible bee. I had lots of blossoms — but no cucumbers.)

Outside, where I mow my roof, you can see the solar panels that are responsible for the heat and natural light at the back of our home. Our partner's children often join ours for a game of football on the roof. But down there you are not bothered at all.

The experience of living underground, in my protected environment, has been a truly magnificent one. We found that the children's allergies are just not the problem that they were above ground, and I'll tell you, my arthritis doesn't hurt down there.

Oh, I suppose I was a skeptic at first. While most men can look at these on paper and determine that they are structurally sound, economically sound, ecologically sound . . . a woman's view seems to be that underground living will surely be dark, damp, gloomy and a real hardship.

Well. it's just not so.

Living in that mass of concrete, with the sunlight and quiet . . . with the earth wrapped around you, the feelings of security, serenity and peace are almost overwhelming to me.

High winds are no longer a threat to my family. During the blizzard this winter most people told me they tried to keep their homes up to 68 degrees. We kept ours down to 72.

To a woman, light, warmth and security are some of the most important things in her life. She is greatly affected by the four walls that surround her.

If you ask me, the very best place to find those IMPORTANT things is underground.